BARLEY FOR FOOD
AND HEALTH

On a Seed

"This was the goal of the leaf
and the root,
for this did the blossom burn
its hour.
This little grain is the
ultimate fruit,
This is the awesome vessel of
power.

For this is the source of the
root and the bud.
World unto world
remolded.
This is the seed, compact of
GOD,
Wherein all mystery is
unfolded."

G. S. Galbraith

BARLEY FOR FOOD AND HEALTH

Science, Technology, and Products

Rosemary K. Newman
C. Walter Newman
Professors Emeritus
College of Agriculture
Montana State University
Bozeman, Montana

WILEY

A JOHN WILEY & SONS, INC., PUBLICATION

For general information on our other products and services or for technical support, please contact
our Customer Care Department within the United States at (800) 762-2974, outside the United
States at (317) 572-3993 or fax (317) 572-4002.

Wiley also publishes its books in a variety of electronic formats. Some content that appears in print
may not be available in electronic formats. For more information about Wiley products, visit our
web site at www.wiley.com.

Library of Congress Cataloging-in-Publication Data:

Newman, Rosemary K.
   Barley for food and health : science, technology, and products / Rosemary K.
Newman and C. Walter Newman.
        p. cm.
      Includes index.
      ISBN 978-0-470-10249-7 (cloth)
      1.   Barley. 2.   Cookery (Barley). I. Newman, C. Walter. II. Title.
      SB191.B2N49 2008
      641.3'316–dc22

                                                                                2008002715

Printed in the United States of America

10 9 8 7 6 5 4 3 2 1

Robert F. Eslick
December 24, 1916–January 19, 1990

This book is dedicated to the memory of Professor Robert F. Eslick, scientist, teacher, mentor, and friend. Professor Eslick was better known as Bob or "Barley Bob" because of his dedication and love of barley. Bob joined the Plant and Soil Science Department at Montana State University in 1946, retiring as Professor Emeritus in 1983. There have been a number of very special people that we have come to know either personally or through their work with barley, but Bob was elite in an elite group. Bob was a great leader, although he was extremely modest about his work and accomplishments. His influence on the barley industry was of an importance and magnitude that continues to this day. He was recognized among plant scientists for his studies on the genetics of barley, contributing especially to the development of new genetic combinations of hulless barley lines with high levels of soluble dietary fiber and β-glucans. He was also a pioneer investigator in barley genetics related to virus disease control, malting barley variety development, feed barley quality, and numerous environmental factors, such as drought, that affect barley production. His contributions were not limited to North American barley, but extended to all continents where barley is grown. Bob's most lasting influence on the barley industry is through his former students, who continue to breed better barleys. Bob was extremely confident that through genetics, improving barley quality was limited only to knowledge of the barley genome and the inspiration of the scientist. I shall always remember his admonition to food and nutrition scientists: "Tell me what you want in a kernel of barley and I will make it be so."

CONTENTS

PREFACE

This book is about the use and potential of barley in food products. As nutritionists, we have often been questioned about our passion for barley food, in that other cereals are produced in abundance, are well accepted as foodstuffs, and provide many of the same essential nutrients as barley. Barley has so much to offer as a source of nutrients and potential nutraceuticals, and as a flavor and texture ingredient products. Information presented in this book will hopefully encourage greater use of barley by commercial food processing companies to produce nutritionally sound and truly health-promoting food products for consumers. We hope to increase the awareness of professionals and students of food science and nutrition of the extensive research that has been reported on barley foods.

The relationship of man and barley goes back to pre-biblical literature. In the histories of most of the early civilizations that cultivated barley, there were almost always references relating to health and well being aspects such as increased stamina, strength, and healing as well as religious and spiritual values attached to this grain. Barley, along with Einkorn and Emmer wheat, flax, and legumes such as lentils, peas, and chickpeas evolved from wild plants, and became domesticated crops purposely selected and used for human sustenance. Domestication and subsequent cultivation of barley and other crops for food were essential elements in the success of human survival and growth. Archaeological evidence indicates that barley kernels grown in ancient times are not too different from modern day barley, exhibiting common features such as hulled and hulless two-rowed and six-rowed barley.

In the chapter on barley taxonomy, morphology, and anatomy we present the similarities of barley with other cereals while illustrating the unique features of the plant. In the following chapters we discuss current progress in barley breeding and the development of transgenic barley, the major genes in the barley genome that influence nutrient composition, processing and product composition. Genetically imposed influences on barley grain composition have a profound influence on product development and nutritional efficacy of the products. Reported food science research provides practical information for new product development using barley, what has been tested, and what will and what won't work. The chapter on health benefits of barley reviews evidence-based data in relevant health and wellness areas. This chapter is perhaps the real "meat" of the book, showing how the unique components of the barley kernel can have profound beneficial effects on the health and well being of consumers.

In the concluding two chapters we provide some insights on the current status of global production and utilization of barley and a few selected traditional recipes from around the world. Contact listings for sources of barley and resource organizations are given in the appendix, along with glossaries of botanical, food and nutrition, and currently accepted barley terminology.

ACKNOWLEDGMENTS

We recognize the Montana Agricultural Experiment Station and the Montana Wheat and Barley Committee for support of our barley research programs at Montana State University. A sabbatical leave to the Agricultural University in Uppsala, Sweden, sponsored by Professors Sigvard Thomke and Per Åman was the initiation of our long-term determination to promote barley as a food grain. We are grateful to our friends and colleagues who have encouraged our work and encouraged us to persist in staying current with barley food development. The many barley researchers who pursue nutritional and product development continue to support our goals through their dedicated work, and it is reports of their work that form the basis of this book. There are far too many barley workers to name individuals who have inspired us. We especially appreciate the camaraderie that has long existed among those who share in studying this fascinating grain.

We thank the following people generously spent time reviewing and editing chapters: Phil Bregitzer, Dale Clark, Peggy Lemaux, Charles McGuire, Graeme McIntosh, Birger Svihus, and Steve Ullrich. Traditional recipes were generously shared with us by Hannu Ahoka, Bung-Kee Baik, Lars Munck, and Tatiyana Shamliyan. We are indebted to Milana Lazetich for her cheerful assistance with manuscript preparation, to Angioline Loredo for her support, and to our editor Jonathan Rose, who unfailingly supplied encouragement and technical assistance.

1 Barley History: Relationship of Humans and Barley Through the Ages

INTRODUCTION

There is considerable historical and archaeological evidence documenting the role of barley as a sustaining food source in the evolution of humankind. Indeed, it was one of the most important food grains from ancient times until about the beginning of the twentieth century. Additionally, alcoholic beverages of various types and fermented foods prepared from barley are commonly referred to in the ancient literature. As other food grains (e.g., wheat, rye, and oats) became more abundant, barley was relegated to the status of "poor man's bread" (Zohary and Hopf 1988). However, modern consumer interest in nutrition and health may help restore barley's status as a significant component in the human diet.

In this chapter we provide some historical perspectives on the origin, domestication, and early food uses of barley up to somewhat recent times. Our knowledge of barley's prehistory comes in large part from archaeological studies of ancient civilizations. Along with archaeological and historical evidence, scientists have used genetics, biochemical, and morphological data to follow the evolution of barley from a wild plant to a domesticated (cultivated) crop. It is generally accepted as fact that the transformation of "wild barley" into a cultivated crop occurred over many millennia (Zohary and Hopf 1988). Fragile ears (spike), a genetic characteristic of wild barley, made it difficult to harvest seeds, as the kernels shatter at maturity. It has been postulated that natural mutations in wild barley produced plants with less fragile ears having larger and more abundant seeds that were naturally preferred and selected for food by hunter-gatherers. One may conjecture that agriculture began when seeds from these plants were planted either accidentally or intentionally, producing a "barley crop."

The first barley foods were probably quite simple. The kernels were probably eaten raw until it was discovered that removing the hulls of hulled types followed

Barley for Food and Health: Science, Technology, and Products,
By Rosemary K. Newman and C. Walter Newman
Copyright © 2008 John Wiley & Sons, Inc.

by soaking and/or cooking in some manner enhanced the texture and flavor. It is also logical to surmise that this may have been when early humans learned about fermentation and how to produce alcoholic beverages.

DOMESTICATION AND USE OF BARLEY FOR FOOD

The Origin of Cultivated Barley

There is some speculation, but it is believed by most authorities that the ancestor of modern barley (*Hordeum vulgare* L.) is identical in most respects to present-day *Hordeum spontaneum* C. Koch. This species is still found in abundance in many parts of Asia and North Africa (Harlan and Zohary 1966; Zohary 1969; Harlan 1978; Molina-Cano and Conde 1980; Xu 1982; Zohary and Hopf 1988; Nevo 1992). *H. vulgare* and *H. spontaneum* are interfertile and differ primarily in the attachment of the kernel to the spike; the latter having a brittle rachis that allows the kernels to shatter at maturity. Archaeologists and other scientists who have attempted to reveal more of the historical development of humankind and human attempts at barley agriculture do not conclusively agree upon the exact site(s) of where these events occurred.

The theory that barley was first domesticated in the Fertile Crescent in the Near East, which spans present-day Israel, northern Syria, southern Turkey, eastern Iraq, and western Iran (Harlan 1978), has been widely accepted but not without controversy. As in many controversies of this nature, there are opposing arguments to this theory. A noted Russian agronomist, N. I. Vavilov proposed that barley originated in two separate centers: one in the mountains of Ethiopia and the second in eastern Asia bordering to the north on present-day Tibet and Nepal and south into India in the subcontinent (Vavilov 1926). In both the Ethiopian highlands and the vast area of Asia proposed by Vavilov, there is an abundance of evidence of early barley culture (Harlan 1978; Molina-Cano et al. 2002). Vavilov's conclusion in 1926 was based on the large diversity of morphological types of cultivated barley that exist in these regions. In a later publication Vavilov indicated that barley was unlikely to have been domesticated in an area (Ethiopia) where the wild ancestor did not exist (Vavilov 1940). Although Ethiopia is widely recognized as a center of extensive genetic diversity in barley types, the area was not seriously considered as a center of origin; however, a study by Bekele (1983) indicated this as a possibility. In support of Bekele's study was the significant revelation that the World Barley Collection at Aberdeen, Idaho contains *H. spontaneum* entries from Ethiopia (Molina-Cano et al. 2002).

The suggestion of barley's possible domestication in Tibet has received support in the discovery of *H. spontaneum* in the Qinghai-Xizang plateau of Tibet (Xu 1982). This finding revived the controversy of cultivated barley's origin in the Far East. Abundant evidence as cited by Molina-Cano et al. (2002) indicates that the East Asian and Indian wild forms of barley are distinctly different from the Near Eastern forms in morphological and biochemical characteristics but have the brittle rachis characteristic of *H. spontaneum*. This evidence strongly suggests that domestication of wild barley occurred in both the Near and Far

East, although domestication in the latter may have occurred considerably more recently (Xu 1982). Furthermore, the geographical distribution of wild barley in North Africa has been extended beyond the boundaries suggested in earlier studies (Harlan 1978) with the finding of *H. spontaneum* plants in southern Morocco (Molina-Cano and Conde 1980). As with the Tibetan barley plants, those found in Morocco were different in many morphological characteristics from those found in Afghanistan, Iraq, Israel, and Libya (Molina-Cano et al. 2002). These authors also presented evidence that suggested the possible existence of wild barley populations in the Iberian Peninsula in Neolithic times. Thus, in light of the evidence, it can be surmised at this time that *H. spontaneum* was present in a vast region in ancient times, beginning in the western Mediterranean region, spanning North Africa, and extending into western, eastern, and southern Asia. The presence of wild barley in these areas indicates that it was available for use and domestication by early people.

Although Harlan (1978) felt very strongly in favor of the Fertile Crescent as the true center of the origin of cultivated barley, evidence gathered and presented over the past 20 years suggests a hypothesis for a multicentric origin for barley (Molina-Cano et al. 2002). Exactly where cultivated barley originated is probably academic, as the important fact that barley was an original food utilized by humans and was vital in the development of many civilizations. The most recent discovery site of barley remnants in a prehistoric setting is only a few thousand years old, which considering the eons of human development is a very short period, and there is so much that is truly unknown and hidden in the veil of ancient history.

The Fertile Crescent

Ancient texts from many cultures in Asia, North Africa, and Europe refer to barley as an important dietary constituent. Identifiable whole seeds and remnants of ground seeds of barley and other cereals have been found in numerous archeological sites that predate writing. Seeds of wild barley were found in a prehistoric camp recently excavated on the southwestern shore of the Sea of Galilee in Israel (Nadel et al. 2004). The age of the camp as determined by ^{14}C measurements is about 23,000 years, which makes this the oldest known site of barley use by humans. Until this discovery, the earliest remains of barley were found in archaeological sites at Wadi Kubbanyia near Aswan in southern Egypt (Wendorf et al. 1979). The Egyptian sites are typical Late Paleolithic and were firmly dated between 18,000 and 17,000 years ago. Several well-preserved chemically carbonized barley grains were recovered at these sites, some of which had retained intact cell structure. Kernel sizes and shapes resembled both wild and modern-day cultivated barleys.

More recent evidence of barley use by ancient people in the Fertile Crescent was dated to approximately 10,000 years ago. This evidence was found in archaeological sites from the Bus Mordeh phase of Ali Kosh, near Deh Luran in Iran and Tell Mureybat in Syria. Wild wheat was also found there, although as in other sites in Syria, Palestine, Mesopotamia, and Asia Minor, barley was the more abundant of the two cereals. From available evidence, barley appears to have been grown on a considerable scale by 7000 to 6500 B.C. at Jarmo in the Iraqi piedmont, and large amounts of two-rowed hulled barley remnants have been unearthed at Beiha, north of Petra in southern Jordan. Six-rowed hulless types appeared at Ali Kosh and two Anatolian sites, Hacilar and Catal Huyuk, dating from 7000 to 6000 B.C. (Harlan 1978).

Archaeologists found a clay tablet from ancient Sumer in Lower Mesopotamia dating about 2700 B.C., which gave a prescription for a poultice that included dried powdered herbs and fruit blended with barley ale and oil. The Sumerian scribes also described the correct method of planting barley. Similarly, a small fragment of pottery with cuneiform script dating about 1700 B.C. was found at Nippur (Egypt) describing recommended irrigation practices for growing barley and the deleterious effect of too much moisture (Kramer 1959).

Several jars were discovered at two excavation sites near the ancient city of Kish in the Tigris and Euphrates river lowlands that contained preserved kernels of barley. These samples were dated in the early Sumerian period, and contemporary with predynastic Egypt, about 3500 B.C. (Hill 1937). Bishop (1936) cites evidence of both barley and wheat being grown in Turkestan in the third millennium B.C. The basic foods of the Sumerian diet were barley, wheat, millet, lentils, pulses, beans, onions, garlic, and leeks. Sumerians were also fond of alcoholic beverages; they developed eight kinds of ale made from barley, eight from wheat, and three from grain mixtures (Tannahill 1988). Perry (1983) relates a curious ancient recipe from the medieval Arab culture, which involves putting unleavened and unseasoned barley dough into closed containers and allowing it to "rot" (ferment) for 40 days. The dough was dried and ground into meal, then blended with salt, spices, wheat flour, and water to make a liquid condiment called *murri*; when mixed with milk, it was called *kamakh*. Other records of ancient agriculture in Turkestan show that barley was grown there at least as long ago as the third millennium B.C. (Bishop 1936). Considerable evidence indicates that agriculture and barley use spread from southwestern Asia, following Neolithic migrations west into North Africa, north to Europe and east to the valley of the Indus (Weaver 1950; Clark 1967). On the basis of historical evidence we were able to document, it would seem that it is possible that barley did evolve as a cultivated crop first in the Fertile Crescent, but as pointed out earlier, this theory has been challenged with good authority.

North Africa

There is documented evidence that barley was a mainstay food crop in North Africa for several thousand years, extending from Ethiopia and Egypt across the southern coast of the Mediterranean Sea to Morocco and southern Morocco.

Both barley and wheat foodstuffs were utilized by Egyptians during the early Neolithic period, and breads and porridges were common daily fare. In Egyptian literature, barley (*it*) is mentioned as early as the first dynasty and is often referred to as barley of "Upper Egypt" or barley of "Lower Egypt" and as white or red barley. Crude granaries dug into the ground were discovered at a site of Egyptian Neolithic culture that contained quantities of well-preserved barley and wheat kernels. This site was thought to have been inhabited between 5000 and 6000 B.C. The prehistoric barley appeared to be very similar to barley grains grown in Egypt at the time of the discovery. Digesta from the alimentary canal of bodies exhumed in an ancient Egyptian cemetery dating about 6000 years old was confirmed to contain barley but no wheat (Jackson 1933).

Historical records from Egypt show abundant evidence of the high esteem in which barley was held. Barley heads appeared on many Egyptian coins, and written records from the fifth (ca. 2440 B.C.), seventh (ca. 1800 B.C.), and seventeenth (ca. 1680 B.C.) dynasties indicate its importance. Barley was intimately entwined in Egyptian religious rites and celebrations, being used as an offering to their gods, in funerals, and even becoming a part of Egyptian legends. Ancient Egyptian records proclaimed barley as a gift of the goddess Isis, and germinated barley kernels symbolized the resurrection of the goddess Osiris. Apart from these uses and its use as food in general, barley was used in brewing, as a medium of exchange, and in therapeutic applications (Weaver 1950). Although barley gained its favor primarily as a staple food, probably as porridge or bread, it was also used in making beer or a beverage called "barley wine." In this period an alelike beverage called *haq* was made from "red barley of the Nile" (Harlan 1978). Barley was used widely in medicinal applications in ancient Egypt and was prescribed in different forms for various maladies. Ground barley preparations, usually mixed with oil, were used as purgatives, applied to wounds to decrease the time required for healing, used as anal suppositories, used to remove phlegm from the respiratory tract, used to treat eye diseases, and most impressively, used as a diagnostic agent for pregnancy and to determine the gender of unborn children (Darby et al. 1976). The effectiveness of the latter two uses is unknown.

Ethiopia has a long history of barley cultivation and diverse agroecological and cultural practices dating back as early as 3000 B.C. The nation is renowned for its large number of landrace barleys and traditional agricultural practices. The diversity of barley types found in Ethiopia is probably not exceeded in any other region of comparable size (Harlan 1978; Bekele et al. 2005). The primary use of barley in Ethiopia continues to be for food, with most of the crop used to make the local bread (*injera*). The Oromo, an ethnic Ethiopian people, incorporate barley in their lifestyle following cultural practices that date back thousands of years, where according to Haberland (1963), barley was the only crop grown by these people in ancient times. Among the Oromo and other people in Ethiopia, barley was considered the holiest of all crops. Their songs and sayings often include reference to barley as the "king of grains." People in the highlands of Ethiopia encouraged their children to consume "lots of barley," believing that it made them brave and courageous. Raw or roasted unripe barley was a favorite food of children (Asfaw 1990), a tradition that continues to modern times (Mohammed 1983). In the Oromo society, special systems, practices, and traditions that involved barley foods and beverages have been maintained for at least the last 400 years. Barley porridge (*merqa* and *kinche*) and fermented and unfermented barley beverages of various consistencies (*tella, zurbegonie, bequre, borde,* and *arequie*) are central to harvest and marriage rituals. The Oromo were originally a nomadic people, and it has been suggested that the development of barley as a crop made an immense contribution in transforming these nomads into a settled farming society, thereby making their life more secure (Mohammed 1983). As noted earlier, a large diversity of barley morphological types is found in Ethiopia, which led to the earlier conclusion that Ethiopian highlands might have been a center of origin of barley.

In Morocco, barley has played a significant role in providing food security throughout the history of that country. Since the beginning of the second millennium B.C., the ruling dynasties in Morocco ensured the security of the populace by maintaining large grain storage facilities, which in effect increased their popularity and secured their positions as rulers. Barley was stored by the rural population, whereas wheat was stored by the urban populations, creating two separate but important systems in safeguarding against famine. Barley and durum wheat were the predominant grains consumed in ancient times (Saidi et al. 2005).

Southern Europe

With movement of civilizations from the Fertile Crescent and the initiation of agricultural trade routes, barley use and cultivation spread throughout the European continent. Evidence of the transfer of barley use from Egypt to southern Europe appears in references made to Egyptian barley use by famous Greeks. Herodotus described barley beer as an important drink among the Egyptians. In one of his visits to Egypt, Pliney the Elder witnessed the use of barley in medical treatments and brought the knowledge to Greece. He is quoted as saying that consuming barley would heal stomach ulcers. Hippocrates is also quoted as saying that Egyptians believed that drinking barley water gave them strength and health (Weaver 1950).

Pliney described recipes for barley *puls*, an oily, highly seasoned paste mixture that was a popular food in Greece (Tannahill 1988). Barley was a common constituent of unleavened bread and porridge eaten by the ancient Greeks. A breadroll claimed by Archestratus in the fifth century B.C. to be the "best barley" was prepared in Lesbos and Thebes by "rounding the dough in a circle and pounding" (by hand) prior to baking. This bread was called *krimnitas* or *chondrinos*, which are terms describing coarsely milled barley (Davidson 1999). A twice-baked barley biscuit called *paximadia*, a favorite Greek food item, was then soaked in broth prior to eating (Kremezi 1997). In more recent times in Greece, a combination of barley and wheat flour was used to produce lighter, crunchier biscuits that did not require soaking before eating. Barley was not fully accepted by all Greeks, however. Aristotle, as well as bakers in the more cosmopolitan communities such as Athens, thought it to be less healthy than wheat (Davidson 1999).

Barley was highly regarded and widely grown by the Romans in many countries which they conquered. The ancient Romans held festivals at planting and harvest times in honor of the goddess Ceres, whom they worshiped as giver of the grain. Offerings of wheat and barley were made to the goddess, "the Cerealia munera," or gifts of Ceres, from which the English word *cereal* is derived (Hill 1937). Barley was the general food of the Roman gladiators, who were called *Hordearii* or "barley men." They believed that barley bread gave them greater strength and increased stamina compared with other foods (Percival 1921). Romans are credited with pioneer advances in dryland farming techniques in North Africa during their occupation of those territories (Hendry 1919). Barley was the grain of choice because of its tolerance to low rainfall in arid lands.

Although barley was considered a respectable, even desirable food in parts of ancient Greece and Italy, Roman soldiers came to look on barley as "punishment rations," even though barley malt was used routinely by the armies for making alcoholic beverages (Davidson 1999). In ancient Rome, bread made from wheat was considered to be more nourishing, more digestible, and in every way superior to barley bread. As in later cultures, barley bread *(panus hordeaceus)* was consumed predominantly by slaves and poor people. After the fall of the Roman Empire, barley bread was considered inferior to rye and wheat breads. However, rich citizens used barley bread as "trenchers" or plates that were edible.

Cultivation of crops, including barley, expanded in a northern and easterly direction from the Aegean area, reaching the Caucasus and Transcaucasus regions during the fifth millennium. Fossilized barley plants that appeared to be hulless types were found in ancient settlements near the village of Ghiljar in the eastern part of the region. In the Caucasus mountain districts above 1700 m, barley was the only grain crop cultivated especially for food by the ancient inhabitants. Most of the barley grown was hulless and was used to make flat cakes and soup. A product capable of being stored, called *ini*, was made by frying hulless barley kernels on a special brazier, then grinding and storing in a large earthenware vessel. *Ini* could be stored for many months, requiring only the addition of water and salt for preparation of dough, which was then rolled into a ball and was ready to eat or could be baked or grilled. Although not necessary and only if available, butter and/or cheese were added to the mixture prior to eating. As in many ancient societies, barley drinks were common among people living in the Caucasus mountain area. One such beverage, *buza*, containing about 4% alcohol, was a traditional drink made from fermented hulless barley cakes and malt (Lisitsina 1984; Omarov 1992).

Southwestern, Central, and Northern Europe

Barley is thought by some historians to have reached Spain in the fourth or fifth millennium B.C., spreading north from there through what is now Switzerland, France, and Germany. However, Molina-Cano et al. (1999; 2002) hypothesized that ancient domestication of barley could have occurred on the Iberian Peninsula and southern France in Neolithic times, independent of genome sources from the Fertile Crescent. Identification of *H. spontaneum* in Morocco and the Iberian Peninsula suggest the possibility that landrace cultivars may have developed in the western Mediterranean area prior to movement of barleys through the trade routes from the Fertile Crescent. There are records of barley's place among Neolithic cultures in many parts of Europe, including Stone Age people in Switzerland (Bishop 1936). According to Davidson (1999), a type of ancient barley bread survives in Jura, a mountainous region of France. This bread, called *bolon* or *boulon*, is prepared in small loaves that are very hard and require soaking in milk or water prior to eating.

Agriculture and cultivated barley reached central and northern Europe during the third millennium B.C. (Körber-Grohme 1987, cited by Fischbeck 2002). It was suggested by Fischbeck that expansion of agriculture also originated from the Aegean region, moving up the river valleys through the Balkans and north into what is now Germany, Poland, and the Baltic countries. Land areas where barley was grown were considered to be less desirable for other crops, yet barley was reported to have thrived under such conditions. Hulless, six-rowed barleys were introduced to the British Isles from the European mainland in about 3000 B.C. (Clark 1967). Coins of early Britons carried pictures of barley bearing the Anglo-Saxon name *barlych* or *baerlic*. Roman records during their occupation of northern Europe indicated that barley was a staple food of the population (Weaver 1950). Bread made from barley and rye flour formed the staple diet of peasants and poorer people in England in the fifteenth century (Kent and Evers 1994). Beyond the Roman period and through the Dark Ages, barley was a dominant food grain throughout Western Europe. It is said to have been the chief bread plant of continental Europe as late as the sixteenth century (Hunt 1904). Oats have long been thought of as the chief element of the Scots' diet, but in fact, at the beginning of the eighteenth century a mixture of barley, peas, and beans was a common food, especially among the poor. Oats were a "rent-paying" crop in Scotland; thus, barley was the cereal eaten by most rural Scots (Gauldie 1981). It was said in the early literature that Scottish land tenants "eat nothing better than barley meal and a few greens boiled together at midday and barley meal porridge at evening and morning."

In the Orkney and Shetland islands off the northern coast of Scotland, a variety of food barley called *Bere* was very popular for milling into flour in Europe for more than 2000 years. Bere is a six-rowed landrace variety adapted to growing on acid soils and to short growing seasons. It is thought to have been introduced to the islands by Norse or Danish invaders in the eight century A.D. (Jarman 1996), although carbonized barley kernels of hulled and hulless types recently excavated in Shetland have been dated to about 1560 B.C. According to Sturtevant (1919), Bere barley, also known as "big barley," was "one of the varieties formerly cultivated in Greece" but once grew wild in the region between the Euphrates and Tigris rivers. The word *Bere* occurred in written form around the thirteenth century and was probably an earlier form or stem of the word for barley and not a proper name (Hannu Ahokas, personal communication). Oats, along with barley, were first grown on the Orkney Islands during the Iron Age. These two grains remained the standard grain crops for many years and were cultivated by the Vikings on many of the Scottish islands (Fenton 1978). Flour milled from Bere barley was traditionally used for making bannocks and similar baked products. Scottish mills were adapted not only for oats but also to grinding of Bere and the pulses, either separately or as a mixture. The resulting mixed meal was made into coarse bread, flat and unleavened, or into porridge. Bere that had been "knockit" (roughly pearled) and left whole was used to make barley broth. Scotch broth or barley broth, a well-known traditional soup, was prepared by boiling beef or lamb with barley, adding a variety of vegetables and a little sugar. The Scottish *mashlum*, a mixture of peas, beans, and Bere barley or oats, was still eaten by all classes until the end of the eighteenth century and by workers after that in some lowland areas. A mixture of peas and barley meal known as "bread meal", was sold by a grain mill at Perth as late as 1837. The popular scones served at tea throughout the United Kingdom were made originally from Bere barley meal or a mixture of Bere barley meal and oat meal. *Scone*, a Scottish word, is perhaps derived from *schoonbrot* or *sconbrot*, meaning white bread (Gauldie 1981). A water-powered mill near Kirkland in the Orkneys produces Bere barley meal in the traditional way today.

According to Mikelsen (1979), hulless barley and einkorn were introduced into Norway between 2000 and 1700 B.C. Of the two grains, barley was far more commonly grown, probably because it was more winter hardy. On the island of Senja in Troms in the far northern part of Norway, a porridge called *vassgraut* ("water porridge") was a common food made by adding ground barley to boiling water. In the Viking era, "ash bread" was prepared by baking barley dough in hot ashes. In addition, it was common to bake barley flatbread on a type of griddle over fires.

Barley was grown extensively in Scandinavia during the Bronze Age (around 2000 B.C.), as wheat of that era was more difficult to grow in the prevailing cold climate. During the Bronze Age, barley became the major cereal for food in Scandinavia, a tradition that continued until early in the twentieth century in many parts of northern Europe. A common diet at the beginning of the twentieth century in Lunnede (Denmark) on the island of Fyn included porridge of barley grits cooked in milk or beer in the morning; meat broth with abraded barley at noon; and barley grits cooked in sufficient amounts to provide for next day's breakfast as well in the evening (Munck 1977).

Perhaps the first clinical trial with barley was reported by L. M. Hindhede in his book *Fuldkommen Sundhed og Vejen Dertil*, in which two adult men were fed a diet based on barley grits, margarine, and sugar for 180- and 120-day periods, respectively. The first subject was involved in heavy manual labor and lost 3.5 kg of weight during the first 36 days of the 180-day period but was otherwise in good health at the end of the study. The second subject, who was in poor health initially, complaining of chronic stomach problems, gained 9 kg in the 120-day period and made the following proclamation: "I have now lived on barley porridge for the last four months. As well as having completely recovered my health, I have gained 9 kilos" (Munck 1977).

Flatbreads made from barley meal were common in Sweden, and the loaves were dried and kept up to six months as a staple food. In the Faroe Islands, dough balls made with milled barley were first placed on an open fire to form a crust, and the baking was completed by placing the crusted barley dough balls in warm ashes. This ancient type of bread, called *drylur*, is not unlike the "ash bread" made from barley flour and water in ancient Norway. Abraded barley kernels and barley grits were used in many ways, including in soups, porridges, meat blends, sausages, and blood mixtures, and in many instances were blended with legumes and other cereals. In Dalarne, Sweden, pea meal and oats were commonly blended with barley meal for baking. Sour beer was often used in food preparation, especially in baking barley or mixed-grain bread (Munck 1977).

East Asia

As was the case with the spread and/or development of barley into North Africa, the Mediterranean region, and Europe, barley cultivation is thought to have moved rapidly eastward from the Fertile Crescent through the trade routes into many parts of Asia, reaching Tibet, China, Japan, and India in the second and third millennia B.C. (Davidson 1999). However, on the basis of genetic and biochemical data presented and summarized by Molina-Cano et al. (2002), it is possible that barley was also domesticated in parts of eastern Asia, particularly in Tibet.

Regardless of barley's origin in Asia, barley has been and continues to be a mainstay in the diet of the Tibetan people. Nyima Tashi, a faculty member at the Tibet Academy of Science in Lhasa, described the long history of the use of hulless barley in food products in Tibet, including various types of cakes, porridges, soups, and snack foods. These products are still prepared using various combinations of roasted hulless barley flour, butter, cheese, sugar, milk, fruits, and meats of different kinds (Tashi 2005). Tashi used the word *tsangpa* to describe roasted grain flour that is prepared by adding cleaned kernels of grain to fine sand previously heated to 100 to 150°C for 2 to 3 minutes and then sieving out the sand. The roasted kernels, called *yue*, are ground into fine flour called *tsangpa*. Shelton (1921) and Ames et al. (2006) described a flour product made from hulless barley roasted and ground in a similar manner as *tsampa*. (We surmise that this is a different English spelling for the same product, due to the similarities in the descriptions of preparation.) Both Shelton and Ames were privileged to have spent time in Tibet, allowing them personally to witness the preparation and use of *tsampa*.

For several thousand years the Tibetan diet consisted mainly of two food items, *tsampa* and yak *butter tea*. Shelton (1921) described *tsampa* preparation as follows: Hulless barley was parched, ground into very fine flour, and made into flat cakes. Butter tea was made from a strong Chinese tea that was strained into a churn to which varying amounts of "more or less stale" yak butter and salt were added. The mixture was then churned into an emulsion. After drinking some of the tea emulsion, the *tsampa* flour was added, kneaded into doughlike balls called *ba*, and eaten. As one would expect, alternative procedures of making *tsampa* have been reported as well, but the basic components are not different. The roasted barley flour was sometimes placed in a bowl or other container first and tea added along with the yak butter. A portion of the tea was drunk, and then the mixture was kneaded with fingers into a doughy paste, formed into small rolls or balls. *Tsampa* could be taken on journeys and eaten dry or with some type of liquid, such as water or milk (Shelton 1921). It is interesting to note the similarity of Tibetan *tsampa* and *ini* made by people who lived in the Caucasus mountain region. *Tsampa*, prepared in much the same way as in early times using native hulless barley, still makes up a substantial part of the Tibetan diet (Tashi 2005; Ames et al. 2006). *Tsampa* is considered a convenience food and is often used by Sherpas, nomads, and other travelers, as in the past. *Tsampa* is used in ways other than food, including as a remedy for toothaches and sore areas. *Tsampa* is traditionally incorporated in certain religious occasions, wedding and birthday celebrations, and the New Year festival. Tossing *tsampa* foods such as the doughlike balls into the air is a symbolic gesture commonly done at various religious and family celebrations in Tibet (Tashi 2005; Wikipedia 2007).

An important and early center of agriculture in the subcontinent of India was in the Indus Valley, now mainly Pakistan. Wheat and barley were staple foods of the Harappan civilization, which flourished in this general area from about 3200 to 2200 B.C. Ajgaonker (1972) described how ancient Indian physicians effectively stabilized type 2 diabetes some 2400 years ago. The treatment was remarkably simple and not really different from recommendations that are given to people with diabetes today (i.e., lose weight, change diet, and increase exercise). In the case of diet, the major changes were reduced caloric intake and substitution of barley for white rice.

Barley has been grown in Korea for many years and in the southern part of the peninsula as a rotation crop with rice, the latter planted in the summer season, with barley planted in the winter season. It is believed that barley was first cultivated as a crop in Korea around 100 B.C (Bae 1979). Although rice is the favorite cereal of Koreans, barley was used as an extender in many rice recipes, especially during periods when rice was in short supply. Various milling procedures have been used in the past, such as splitting the kernel at the crease to decrease cooking time and making the barley kernel closer in size to a grain of rice. In the Korean language, cooked cereal, which is usually rice, is called *bob* and mixed *bob* is prepared by cooking rice and adding precooked barley. *Bori* is the Korean word for barley, and when 100% of barley is used for *bob*, the dish is called *kkong bori-bob*. Barley malt has been used to prepare a traditional sweet drink, and in some instances barley has been used in the preparation of fermented soybean paste as well as hot pepper paste in Asia (B.K. Baik, personal communication).

Barley tea has a long history in Asia and is still a popular drink in many parts of the continent, including Korea, China, Japan, India, and especially, Tibet. Barley tea is prepared from roasted barley kernels that are steeped, making a mild nonalcoholic drink that is consumed both hot and cold, with and without meals, in place of water. In the past, kernels were roasted at home for use in tea. Today, however, roasted kernels for making barley tea are available in many food stores in modern cities and towns. There are many anecdotal references to the medicinal value of barley tea for young and old throughout the barley literature.

In a section titled "Recipes for the Sick" in *The Rumsford Complete Cookbook* (Wallace 1930), a recipe for barley water for the sick is given as follows: 2 Tbsp pearl barley, 1 qt cold water $\frac{1}{3}$ tsp salt, juice of one-half a lemon, and a little sugar; soak the washed barley, add salt, and cook for about 3 hours; strain through cheesecloth, flavor with the lemon juice and sugar as desired.

North and South America

The story of barley in the New World is a recent and brief fragment of the history of the use of this grain by humankind (Weaver 1950). Columbus brought barley to the North American continent on his second voyage, in 1494 (Thacher 1903). The original introduction site was not conducive to barley culture and there were no further reports of production in that area, although there are records of barley crops in Mexico in the sixteenth century (Capettini 2005). Later, there were two additional pathways through which barley was introduced more successfully into North America. Barley was brought to the east coast colonies from England at the turn of the seventeenth century and into the southwest during the Spanish mission movement (Weaver 1950; Wiebe 1978). Barley was probably planted for the first time on Martha's Vineyard and Elizabeth Island off the coast of present-day Massachusetts in 1602. Four years later, barley was planted at Port Royal, Nova Scotia and in Champlain's garden in Quebec as early as 1610. Barley was also grown by Dutch colonists in "New Netherland," and samples were shipped to Holland in 1626. Colonists of the London Company grew barley in Virginia in 1611, and it became an important crop by the middle of the seventeenth century. In the colonial days, barley was grown principally for beer production, and the varieties cultivated were those common to the home region of the colonist. Although barley production flourished and spread westward, the early varieties introduced from northern Europe were not well adapted to the soil and climate of the eastern United States and Canada. As barley moved west in North America, production was centered close to populated areas to provide raw material for breweries (Weaver 1950). One of several such enterprises, the Manhattan Malting Company, was established by Henry Altenbrand and Jacob Rupert in 1891 in the Rocky Mountain region of western Montana, near the present town of Manhattan. Henry Altenbrand was president of the New York and Brooklyn Malting Company, and Jacob Rupert was the owner of the New York Yankees baseball team. In celebration of the newly founded brewery and barley-growing company, Altenbrand had the town name officially changed from Moreland to Manhattan. This operation was highly successful, contributing to the financial support and population growth of the area. The Manhattan Malting Company, as well as many similar enterprises in the western United States, was doomed by Prohibition in 1916. However, the initial start of barley production in the Manhattan, Montana area continued as demand for barley for animal feed grew and farmers diversified into other crops (Strahn 2006). Thus, the popularity of barley beverages played a significant role in settling many parts of the Old West of the North American continent. Barley spread north as well as west from the original east coast introductions. There is a report of barley cultivation as far north as Fort Yukon, Alaska in the late nineteenth century (Dall 1897).

The early history of barley's establishment in North America is not confined to the east coast colonies. The Spanish brought barley into the area that is now Mexico, resulting in eventual distribution to South America, the American southwest, and California in the sixteenth and seventeenth centuries. In most instances, barley in these areas was used primarily for animal feed and secondly for beer production. Barleys that were introduced to this part of North America by the Spanish explorers were very well adapted to the soil and climate, having been developed under similar ecological conditions in North Africa and in regions of southern Europe. Thus, barley was established on both sides of the North American continent, moving from the coastal areas into the interior as the population grew and moved into new territories (Weaver 1950). Barleys that were adapted to the high mountains of Africa prospered in similar environments of the New World, and those adapted to the hot, dry coastal areas of the Mediterranean Sea prospered in Mexico, the southwestern desert country of the United States, and California (Harlan 1957).

Simultaneous to the introduction of barley into Mexico, there are records of barley crops in the sixteenth century in Argentina (1583), Peru (1531), Chile (1556), and Brazil (1583). In 1914, two German scientists, Alberto Boerger and Enrique Klein, initiated the first formal barley program in Latin America at the La Estanzuela Experiment Station in western Uruguay. Local landrace barleys were used as gene sources for developing new, adapted, and improved feed and malting lines. Klein later moved to Argentina, where he continued his barley breeding programs using germplasm he brought from Uruguay (Capettini 2005). Harlan (1957) described barley grown in Peru at the beginning of the twentieth century as very much like the "coarse" barley that was introduced into South America by Spaniards. He suggested that these barleys originated from high mountain regions of southern Europe or North Africa, as they were being produced at elevations up to 4000 m in Peru. Although the native people of Peru had proven very adept at breeding other food plants, Harlan (1957) observed that there was very little evidence of improvements made in barley.

There is little or no conclusive historical evidence that barley was used for food to any great extent other than as a beverage during these early years in North or South America. It can be deduced with some accuracy that most of the barley crop produced during the settling of the New World by people of European origin was used primarily for malt production and animal feed. The same pattern continues somewhat today in the United States and the southern cone of South America, although there are local areas in parts of South America, especially Ecuador, where barley represents a major food source in rural areas (Villacrés and Rivadeneira 2005).

Contacts initiated in 1492 between the Old and New Worlds brought profound and lasting change in human populations and plant life to the entire planet. Barley was only one of many biological entities introduced into the Americas with the migration of Europeans. The westward movement of barley and other crops to the New World was accompanied by the introduction of crops such as maize and potatoes into European and African agriculture (Crosby 2003).

SUMMARY

In this chapter we have attempted to present a logical progression and development of barley as a food along with developing civilizations through the ages. Cereal grains have long been noted for their essential contributions to human survival throughout history. Barley, in particular, was a dietary mainstay of ancient civilizations and continued to be an important dietary constituent of working-class people in Europe until the end of the nineteenth century. On the basis of archaeological evidence, it can be speculated that the evolution of barley foods and beverages paralleled the early development of the human race. Locating the origin of cultivated barley has not been without controversy. The most prominent and accepted theory of origin is in the Fertile Crescent. There is, however, compelling evidence of the possibilities of multicenters of origin of barley, initiating in the Iberian Peninsula, extending across North Africa, southwestern Asia, and into eastern and southern Asia. Throughout historical and archaeological reports, barley is referred to as a source of health, strength, and stamina for athletes and persons involved in hard manual labor. The health benefits and medical aspects of barley foods are stressed in ancient Arabian, Chinese, Egyptian, Ethiopian, and Greek literature. These same beneficial properties of barley foods that were recognized by the ancients were also touted by more recent civilizations, extending from Asia to Europe. In almost every culture through past ages, barley foods are described as having almost mystical properties. Barley cultivation moved with advancing civilizations through Europe and into the New World. In many instances barley was grown in areas and under conditions that were not suitable for other crops, providing a source of nutrients to the less fortunate section of the population. Barley lost favor as a food grain due primarily to improved conditions of the farming classes and the growth and development of the wheat industry. Wheat bread and wheat-based breakfast cereal products have replaced many of the bakery markets for rye, oats, and barley because of texture, taste, appearance, and increased availability.

REFERENCES

Ajgaonker, S. S. 1972. Diabetes mellitus as seen in the ancient Ayurvedic medicine. Pages 13–20 in: *Insulin and Metabolism*, J. S. Bajaj, ed. Association of India, Bombay; India.

Ames, N., Rhymer, C., Rossnagel, B., Therrien, M., Ryland, D., Dua, S., and Ross, K. 2006. Utilization of diverse hulless barley properties to maximize food product quality. *Cereal Foods World* 51:23–26.

Asfaw, Z. 1990. An ethnobotanical study of barley in the central highlands of Ethiopia. *Biol. Zentbl.* 109:51–62.

Bae, S. H. 1979. Barley breeding in Korea. Pages 26–43 in: *Proc. Joint Barley Utilization Seminar*. Korea Science and Engineering Foundation, Suweon, Korea.

Bekele, E. 1983. A differential rate of regional distribution of barley flavonoid patterns in Ethiopia, and a view on the center of origin of barley. *Hereditas* 98:269–280.

Bekele, B, Alemayehu, F., and Lakew, B. 2005. Food barley in Ethiopia. Pages 53–82 in: *Food Barley—Importance, Uses and Local Knowledge: Proc. International Workshop on Food Barley Improvement, Jan. 2002*. S. Grando and H. G. Macpherson, eds. ICARDA, Aleppo, Syria.

Bishop, C. W. 1936. Origin and early diffusion of the traction-plough. *Antiquity* 10: 261–278.

Capettini, F. 2005. Barley in Latin America. Pages 121–126 in: *Food Barley—Importance, Uses and Local Knowledge: Proc. International Workshop on Food Barley Improvement, Jan. 2002*. S. Grando and H. G. Macpherson, eds. ICARDA, Aleppo, Syria.

Clark, H. H. 1967. The origin and early history of the cultivated barleys. Pages 1–18 in: *The Agricultural History Review*. British Agricultural History Society, London.

Crosby, A. W. 2003. *The Columbian Exchange: Biological and Cultural Consequences of 1492*. Praeger Publishers, Westport, CT.

Dall, W. H. 1897. *Alaska and Its Resources*. Lee and Shepard, Boston.

Darby, W. J., Ghalioungui, H., and Grivetti, L. 1976. *Food: The Gift of Osiris*. Academic Press, New York.

Davidson, A. 1999. *The Oxford Companion to Food*. Oxford University Press, Oxford.

Fenton, A. 1978. Grain types and trade. Pages 332–336 in: *The Northern Isles: Orkney and Shetland*. Tuckwell Press, East Lothaian, Scotland, UK.

Fischbeck, G. 2002. Contribution of barley to agriculture: a brief overview. Pages 1–14 in: *Barley Science: Recent Advances from Molecular Biology to Agronomy of Yield and Quality*. G. A. Slafer, J. L. Molina-Cano, R. Savin, J. L. Araus, and I. Romagosa, eds. Haworth Press, Binghamton, NY.

Gauldie, E. 1981. Diet: the product of the mill. Pages 1–21 in: *The Scottish Country Miller 1700–1900: A History of Water-Powered Meal Milling in Scotland*. John Donald Publishers, London.

Haberland, E. 1963. *Völker Süd-Aethiopiens*, vol. 3; *Gala Süd-Aethiopiens*. Kohlhammer, Stuttgart, Germany.

Harlan, H. V. 1957. *One Man's Life with Barley: The Memories and Observations of Harry V. Harlan*. Exposition Press, New York.

Harlan, J. R. 1978. On the origin of barley. Pages 10–36 in: *Barley: Origin, Botany Culture, Winter Hardiness, Genetics, Utilization, Pests*. Agriculture Handbook 338. U.S. Department of Agriculture, Washington, DC.

Harlan, J. R., and Zohary, D. 1966. Distribution of wild wheats and barley. *Science* 153:1074–1080.

Hendry, G. W. 1919. Mariout barley, with a discussion of barley culture in California. *Calif. Agric. Exp. Sta. Bull. 312*. University of California, Berkley, CA. (Cited by Weaver, 1950.)

Hill, A. F. 1937. *Economic Botany: A Textbook of Useful Plants and Plant Products*. McGraw-Hill, New York.

Hunt, T. F. 1904. *The Cereals in America*. Orange Judd Company, New York.

Jackson, A. 1933. Egyptian Neolithic barley. *Nature* 131:652.

Jarman, R. J. 1996. Bere barley: a living link with the 8th century. *Plant Var. Seeds* 9:191–196.

Kent, N. L., and Evers, A. D. 1994. *Kent's Technology of Cereals*, 4th ed. Elsevier Science, Oxford.

Köber-Grohne, U. 1987. *Nutzpflanzen in Deutschland*. K. Theiss, Stuttgart, Germany.

Kramer, S. N. 1959. *History Begins at Sumer*. Doubleday Anchor Books, Doubleday and Co., New York.

Kremezi, A. 1997. Paximadia (barley biscuits). In: *Foods on the Move: Oxford Symposium on Food History*, 1996, Totnes, UK. Prospect Books, London.

Lisitsian, G. N. 1984. The Caucasus: a centre of ancient farming in Eurasia. Pages 285–292 in: *Plants and Ancient Man*. W. van Zeist and W. A. Casparie, eds. Balkema, Rotterdam, The Netherlands.

Mikelsen, E. 1979. *Korn er liv*. Statens Kornforretning, Oslo, Norway.

Mohammed, H. 1983. The Oromo of Ethiopia 1500–1850 with special emphasis on the Gibe region. Ph.D. dissertation. University of London, London.

Molina-Cano, J. L., and Conde, J. 1980. *Hordeum spontaneum* C. Koch em. Bacht. collected in southern Morocco. *Barley Genet. Newsl.* 10:44–47.

Molina-Cano, J. L., Moralejo, M., Igartua, E., and Romagosa, I. 1999. Further evidence supporting Morocco as a centre of origin of barley. *Theor. Appl. Genet.* 73:531–536.

Molina-Cano, J. L., Igartua, E., Casas, A-M., and Moralejo, M. 2002. New views on the origin of cultivated barley. Pages 15–29 in: *Barley Science: Recent Advances from Molecular Biology to Agronomy of Yield and Quality*. G. A. Slafer, J. L. Molina-Cano, R. Savin, J. L. Araus, and I. Romagosa, eds. Haworth Press, Binghamton, NY.

Munck, L. 1977. Barley as food in Old Scandinavia especially Denmark. Pages 386–393 in: *Proc. 4th Regional Winter Cereal Workshop: Barley*, vol. II, Amman, Jordan.

Nadel, D, Weiss, E., Simchoni, O., Tsatskin, A., Danin, A., and Kislev, M. 2004. Stone Age hut in Israel yields world's oldest evidence of bedding. *Proc. Natl. Acad. Sci. U.S.A* 101:6821–6826.

Nevo, E. 1992. Origin, evolution, population genetics and resources for breeding of wild barley, *Hordeum spontaneum*, in the Fertile Crescent. Pages 19–43 in: *Barley: Genetics, Biochemistry, Molecular Biology and Biotechnology*. P. R. Shewry, ed. C.A.B. International, Wallingford, UK.

Omarov, D. S. 1992. Barley for food in mountainous Caucasus. Pages 192–200 in: *Barley for Food and Malt: ICC/SCF International* Symposium. Swedish University of Agricultural Sciences, Uppsala, Sweden.

Percival, J. 1921. *The Wheat Plant*. Duckworth, London.

Perry, C. 1983. A nuanced apology to rotted barley. In: *Petits Propos Culinaires*. Prospect Books, London.

Saidi, S., Lemtouni, A., Amri, A., and Moudden, M. 2005. Use of barley grain for food in Morocco. Pages 17–21 in: *Food Barley—Importance, Uses and Local Knowledge: Proc. International Workshop on Food Barley Improvement*, Jan. 2002. S. Grando and H. G. Macpherson, eds. ICARDA, Aleppo, Syria.

Shelton, A. L. 1921. Life among the people of eastern Tibet. *National Geographic*. 295–326.

Strahn, B. D. 2006. Manhattan: the heart of the Gallatin Valley. Pages 11–13 in: At Home, *Bozeman Daily Chronicle*, Aug. 8.

Sturtevant, E. L. 1919. *Sturtevant's Notes on Edible Plants*. N.Y. Department. of Agriculture, 27th Annual Report. U. P. Hedrick, ed. Dover Publications, New York.

Tannahill, R. 1988. *Food in History*, rev. ed. Penguin, London.

Tashi, N. 2005. Food preparation from hull-less barley in Tibet. Pages 115–120 in: *Food Barley—Importance, Uses and Local Knowledge: Proc. International Workshop on Food Barley Improvement*, Jan. 2002. S. Grando and H. G. Macpherson, eds. ICARDA, Aleppo, Syria.

Thacher, J. B. 1903. *Christopher Columbus: His Life, His Work, His Remains*, vol. 2. G. P. Putnam's Sons, New York.

Vavilov, N. I. 1926. *Studies on the Origin of Cultivated Plants*. Institute de Bontanique Appliqué et d'Amelioration des Plants, Leningrad, Russia. (Cited by Weaver 1950; Harlan 1978.)

Vavilov, N. I. 1940. The new systematics of cultivated plants. Pages 549–566 in: *The New Systematics*. Oxford University Press, Oxford, UK.

Villacfes, E., and Rivadeneira, M. 2005. Barley in Ecuador: production, grain quality for consumption and perspectives for improvement. Pages 127–137 in: *Food Barley—Importance Uses and Local Knowledge: Proc. International Workshop on Food Barley Improvement*, Jan. 2002. S. Grando and H. G. Macpherson, eds. ICARDA, Aleppo, Syria.

Wallace, L. H. 1930. Barley water. Page 222 in: *The Rumford Complete Cookbook*. Rumford Co., Rumford, RI.

Weaver, J. C. 1950. *American Barley Production*. Burgess Publishing, Minneapolis, MN.

Wendorf, F., Schild, R., El Hadidi, N., Close, A. E., Kobusiewicz, M., Wieckowska, H., Issawi, B., and Haas, H. 1979. Use of barley in the Egyptian Late Paleolithic. *Science* 28:1341–1347.

Wiebe, G. A. 1978. Introduction of barley into the New World. Pages 1–9 in: *Barley: Origin, Botany, Culture, Winter Hardiness, Genetics, Utilization, Pests*. Agriculture Handbook 338. U.S. Department of Agriculture, Washington, DC.

Wikipedia. 2007. Published online at http://www.wikipedia.org/wiki/Tsampa.

Xu, T. W. 1982. Origin and evolution of cultivated barley in China. *Acta Genet. Sci*. 9:440–446.

Zohary, D. 1969. The progenitors of wheat and barley in relation to domestication and agricultural dispersal in the Old World. Pages 47–66 in: *The Domestication and Exploitation of Plants and Animals*. P. J. Veko and G. W. Dimbleby, eds. Duckworth, London.

Zohary, D., and Hopf, M. 1988. *Domestication of Plants in the Old World: The Origin and Spread of Cultivated Plants in West Asia, Europe and the Nile Valley*. Clarendon Press, Oxford, UK.

2   Barley: Taxonomy, Morphology, and Anatomy

INTRODUCTION

As shown in Chapter 1, barley is among the most ancient of cereal crops, evolving over eons from wild to domesticated plants. A defining characteristic between wild and modern domesticated (cultivated) barley is the toughness of the *rachis* or main axis of the spike. The rachis connects the tiller (stem) to the spike (ear), and the wild barley rachis is brittle, whereas that of cultivated barley is tough. A tough rachis allows one to reap a harvest. In contrast, the brittle rachis in wild species has been referred to as a "reseeding device" designed to preserve the species (Reid and Wiebe 1978). A brittle rachis results in shattering as soon as seeds are ripe, an effective means of seed dispersal under natural conditions, but a distinct disadvantage for harvesting seed unless the plant is harvested prior to ripeness. Morphological and anatomical characteristics such as rachis strength are used to distinguish between the various species within the genus *Hordeum*. Such anatomical characteristics are easily viewed, whereas other characteristics are more difficult to determine without dissection or microscopic evaluation. Although the primary objective of this book is to explain the health benefits of the barley kernel and use of barley in foods, it is helpful to have a basic understanding of the complete plant to fully appreciate the uniqueness, yet similarity to other cereal grains. An understanding of the taxonomy, morphology, and anatomy of crops such as barley is particularly essential to plant breeders. In this chapter we introduce the science of barley taxonomy and briefly describe the basic morphology and anatomy of the plant and kernel.

TAXONOMY

Linnaeus was the first to provide a botanical description of barley in his *Species Plantarium* in 1753 (Bothmer and Jacobsen 1985), a description that has seen

Barley for Food and Health: Science, Technology, and Products,
By Rosemary K. Newman and C. Walter Newman
Copyright © 2008 John Wiley & Sons, Inc.

many modifications in the scientific literature since that time. Descriptions of the genus *Hordeum* given by Åberg and Wiebe (1946, 1948) defined cultivated and wild barley as follows: "Genus *Hordeum*: Spike indeterminate, dense, sometimes flattened, with brittle, less frequently tough awns. Rachis tough or brittle. Spikelets in triplets, single flowered, but sometimes with rudiments of a second floret. Central florets fertile, sessile or nearly so: lateral florets reduced, fertile, male or sexless, sessile or on short rachis. Glumes lanceolate or awnlike. The lemma of the fertile flowers awned, awnleted, awnless, or hooded. The back of the lemma turned from the rachis. Rachilla attached to the kernel. Kernels oblong with ventral crease, caryopsis usually adhering to lemma and palea. Annual or perennial plants." This description, published in the 1946, was followed two years later by the following: "Summer or winter annuals, spikes linear or broadly linear, tough or brittle. The central florets fertile; the lateral ones fertile, male, or sexless. Glumes lanceolate, narrow or wide, projecting into a short awn. The lemma in fertile florets awned, awnleted, awnless, or hooded; the lemma in male or sexless florets without awns or hoods. Kernel weight 20 to 80 mg. The chromosome number in the diploid stage is 14."

Barley is a grass belonging to the family Poaceae, the tribe Triticeae, and the genus *Hordeum*. There are 32 species, for a total of 45 taxa in the genus *Hordeum*, that are separated into four sections (Bothmer 1992), although as many as six sections have been suggested (Reid and Wiebe 1978). The four sections proposed by Bothmer are as follows: *Hordeum, Anisolepis, Critesion*, and *Stenostachys*. The division of the genus into sections puts plants into groups that have similar morphological characteristics, life forms, similarities in ecology, and geographical area of origin. The basic chromosome number of $x = 7$ is represented across the 45 taxa as diploid ($2n = 2x = 14$), tetraploid ($2n = 4x = 28$), and hexaploid ($2n = 6x = 42$). Six species are listed in the section *Hordeum; H. bulbosum, H. murinum* ssp. *glaucum, H. murinum* ssp. *leporinum, H. murinum* ssp. murinum L., *H. vulgare* ssp. *vulgare*, and *H. vulgare* ssp. *spontaneum*. The genomes of *H. vulgare* ssp. *vulgare* (cultivated barley) and *H. vulgare* ssp. *spontaneum* (wild barley) are identical and interfertile (Fedak 1995). A major difference between these two subspecies is the tough rachis of cultivated barley as opposed to the brittle rachis of wild barley. Additionally, *H. vulgare* may have two- or six-rowed spikes, whereas the spikes of *H. spontaneum* are mostly two-rowed. There are three six-rowed *Spontaneum* species listed in the world barley collection at Aberdeen, Idaho. The six-rowed forms in *H. vulgare* are thought to be due to mutations and hybridization (Nilan and Ullrich 1993). The genetic relationship of *H. vulgare* (cultivated barley) to *H. bulbosum* and *H. murinum* is not as close as that with *H. spontaneum*, but all barleys in this group have similar morphological characteristics and evolved in the same geographical region. Based on similarities in isoenzyme patterns, *H. bulbosum* is probably more closely related to *H. vulgare* than is the *H. murinum* species group (Jörgensen 1982).

Most evidence indicates that the immediate ancestor of cultivated barley is the two-rowed wild barley *H. spontaneum*. Harlan (1978) postulated that the true ancestor of cultivated barley was the progenitor of *H. v. spontaneum* and that this ancestor plant is extinct and no longer exists in the modern world. Harlan suggested that the line of development of barley began with the unknown but closely related plant of *H. spontaneum*, which was eventually transformed through mutations into *H. vulgare*. The morphological variability seen in cultivated barley has been ascribed to changes occurring during cultivation over an extended period of time in a wide array of geographical areas and to intense breeding (Wiebe and Reid 1961).

The position of barley within the Poaceae (grass family) is of interest from the evolutionary viewpoint but also reveals the important relationship with other members of the Triticeae tribe, rye (*Secale cereale*) and wheat (*Triticum* spp.). Taxonomic classification of barley not only reveals these relationships but also allows the identification of barley types and varieties from the morphological characteristics of the plant and grain. The chief identifying taxonomic characteristic of *Hordeum* is its one-flower spikelet, three of which alternate on opposite sides of each node of the flat rachis, forming a triplet of spikelets at each node. There are one central and two lateral spikelets. All three spikelets may be fertile (six-rowed) or only the central spikelet is fertile (two-rowed) (Figure 2.1). The taxonomic descriptions of the genus *Hordeum* as presented is a basic introduction to the science. A detailed description of barley taxonomy, morphology, ecology, and distribution of the various species of *Hordeum* may be found in publications by Bothmer and Jacobsen (1985), Reid (1985), Bothmer (1992), and Moral et al. (2002).

FIGURE 2.1   Barley heads. *Left:* front and side view of six-rowed barley. *Right:* front and side view of two-rowed barley. (Courtesy of American Malting Barley Association.)

The focus of this book is limited to cultivated barley, although it should be recognized that wild barley with brittle rachis is harvested for food in certain areas of the world, especially when famine is prevalent. To avoid grain loss from shattering, wild barley is harvested prior to maturity while green and then allowed to dry, thus preserving the kernel (Harlan 1978).

MORPHOLOGY AND ANATOMY

Growth and development begins with germination of the seed. Moral et al. (2002) divided the growth and development of the barley plant into three major phases: vegetative, reproductive, and grain filling. The vegetative phase begins at germination followed by floral initiation characterized by early formation of tillers and leaves. The reproductive phase begins with the formation of the spikelet and ends with pollination of the ovaries in the spikelets. Grain filling begins after pollination with enlargement of endosperm and embryo cell numbers for accumulating dry matter and initial vegetative primorda, respectively. Duration of each phase varies with genotype, geographic area, and agronomic and climatic factors. The basic anatomical parts of the cultivated barley plant are roots, stems, leaves, spikes, spikelets, and kernels (Figure 2.2), which are not unlike the parts of most plants in the Poaceae family.

Roots

Upon absorption of moisture, germination occurs, initiating a series of events beginning with the formation of roots. Roots of the barley plant are simple axial structures having no leaflike organs, nodes, or internodes. Barley has two sets of roots, the primary or seminal roots and the adventitious roots. Primary roots are produced during germination arising from the coleorhiza, usually numbering five to seven. These roots grow outward and downward, branching and forming a fibrous mass, having hair roots for increasing surface area to maximize water and nutrient absorption from the soil. Hair roots are short lived and are being replaced continuously as the root system grows. Adventitious roots, arising out of the crown as tillers develop, tend to be thicker and are less branched than the primary roots. Seedling establishment begins with the downward extension of roots into the soil. The extent of root development is dependent on the type and depth of soil, availability of water and nutrients, and barley genotype.

Tillers

The coleoptile, a sheath enclosing tissue capable of mitosis, produces a "shoot" that breaks through the testa or seed coat and grows up the dorsal side of the kernel under the lemma shortly after roots begin to develop, providing further stability to the seedling (Figure 2.3). The coleoptile encloses and protects the shoot as it grows toward the soil surface. At this point a complex series of

FIGURE 2.2   Parts of barley plant. *Left:* roots, stem, leaves. *Center:* kernel. *Upper right:* spike or head with awns. *Lower right:* Spikelet. (Courtesy of the U.S. Department of Agriculture.)

events leads to emergence of the cotyledon (first leaf) and the main shoot and primary tillers. Once the cotyledon emerges, the coleoptile ceases to elongate. As the young plant grows, this sequence is repeated at other primary tiller sites. Each primary tiller has the potential to produce secondary tillers. The first primary tillers may become nearly as large as the main shoot. The number of tillers per plant is influenced by plant density, genotype, and environmental factors. Tiller length also depends on genotype and environment, and to some extent, plant density. The mature tiller is a cylindrical structure consisting of hollow internodes, separated by solid nodes or joints with transverse septa. In the average plant there are five to seven internodes, although 10 or 11 internodes have been observed. The basal internode is the shortest, and the internode diameter decreases toward the top of the plant. The peduncle is the part of the tiller from the last node to the collar which marks the transition from the tiller into the rachis of the spike. The collar is sometimes referred to as the *basal node* of the rachis. There several types of collars, the most common being closed, V-shaped, open, and modified closed or open. The top of the peduncle just below the spike is referred to as the *neck* and may be straight, curved, or zigzag.

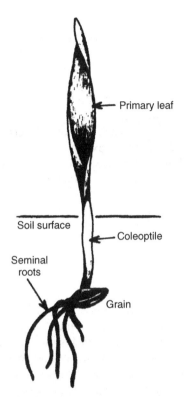

Primary leaf

Soil surface

Coleoptile

Seminal
roots

Grain

FIGURE 2.3   Barley seedling with emerged shoot and primary leaf. (Adapted from Percival 1902.)

Leaves

A barley leaf consists of a sheath, ligule, auricles, and blade. The *blade* is lanceolate-linear or gradually tapered with a prominent middle nerve or rib, flanked by 10 to 12 or more parallel side nerves. The uppermost blade, called the *flag leaf*, is often the smallest leaf, other than the seedling leaf, which is the first leaf to show and is also the smallest. *Foliage leaves* consist of the dermal system, ground tissue, and a vascular system and contain green photosynthetic mesophyll covered by the epidermis on both surfaces. *Stomata*, providing avenues for oxygen intake and carbon dioxide removal, are located on the underside of the leaves. Stomata numbers vary with environment and genotype. Between five and 11 single leaves arise as a semicircular ridge at alternating points on opposite sides of the stem. Leaves will differ in size and shape and position on the stem in different barleys. Leaf size is also influenced by environmental growing conditions, especially day length. The leaf is attached to the stem by the leaf sheaf at the node. The *sheaf* surrounds the stem, being split to the base on the opposite side from the blade with the two edges overlapping each other. The *auricles*, clawlike appendages that clasp the stem, are located at the juncture (top) of the leaf sheath and the blade. The *ligule* is an appendage also located at the juncture of the leaf sheaf and blade extending upward along the stem and varies in size with barley type. The ligule and auricles are absent in some barleys. In addition to providing leaf attachment, the leaf sheaf also provides strength and support to the stem. The angle of attachment of the leaf to the stem varies from erect to drooping, depending on the cultivar.

Because of the chlorophyll, the leaf of normal barley is completely green, but chlorophyll content varies and in extreme cases is so low that the plant is considered an albino. Chlorophyll content is highly correlated with plant survival. The green color of normal barley is intensified by fertilizer, especially nitrogen fertilizer. A red to purple color, due to an anthocyanin pigment, occurs in the sheaths at the base of the plant. The color is enhanced by cold weather. In most barley, the leaf and leaf sheath surfaces are covered by a chalklike substance that creates a waxy appearance. In barleys having less or none of this substance, the plant has a glossy appearance.

Spike

The inflorescence (flower head) of barley, termed the *spike*, is classed as an indeterminate inflorescence because the axis (rachis) does not terminate in a spikelet. The barley spike, located at the tip of the stem neck, consists of varying numbers of spikelets attached at the nodes of a solid, flat, zigzag rachis (Figure 2.4). Barley flowers are complete flowers, containing both the ovule and the anthers and are mostly self- pollinated. Spikelets are single flowered, consisting of two glumes and a floret in which the kernel develops following fertilization. There are three spikelets attached at each node on the rachis, alternating from side to side. All three spikelets are fertile in six-rowed barleys, whereas only the central spikelet is

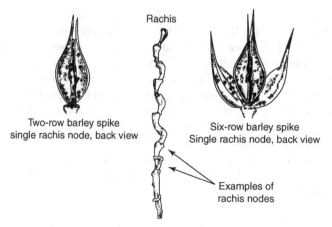

FIGURE 2.4   Spikes and rachies of barley. (From brewingtechniques.com.)

fertile in two-rowed barleys. Some two-rowed barleys have anthers in the lateral florets. In six-rowed barleys, the two lateral rows of kernels on each side of the spike may overlap, giving the appearance of four-rowed barley; however, these barleys are true six-rowed types. Lateral kernels in most six-rowed barleys have a slight twist or bend at the end attached to the spike. (Row type, a genetically controlled characteristic, is discussed in more detail in Chapter 4.) The two glumes, once referred to as *outer glumes*, are linear to lanceolate, terminate in an awn, and are generally equal in size and shape. In each floret the ovary and stamens are enclosed in two flowering glumes, the *lemma* and *palea*, which in the mature kernel become the hulls. The anatomical features of the lemma and palea are similar, but the palea is the smaller of the two. The lower dorsal glume (lemma) slightly overlaps the upper ventral glume (palea). In most cultivated barley, the lemmae extend to long thin awns that are roughly triangular in cross section, but in some barley the awns are short, curly, or may be "hooded" (Figure 2.5). Lemmae that are genetically coded to bear hoods develop like the extended awned forms, but the apex grows more slowly with the appearance of lateral "wings,"

FIGURE 2.5   Beardless and hooded types of barley. *Left to right:* beardless, elevated hoods, and nonelevated hoods. (Courtesy of American Malting Barley Association.)

and the apex is formed into a hollow structure that may or may not have a true awn extension. Barleys having extended awns are often referred to as having *beards* or being bearded. All awns, extended or hoods, contain photosynthetic tissue and stomata. Stomata are also present on the inside of the lemma.

Spacing of the mature kernels on the rachis partially determines the overall shape of the spike. At maturity, spikes may be upright or bent, which is sometimes described as *nodding*, with all degrees of intermediate expression of this characteristic. Length of spike, thus the number of spikelets per spike, is also variable, depending on genotype. Environmental conditions also influence spike length, and length may differ on different tillers on the same plant.

Kernels

About two months following sowing, *anthesis* or shedding of pollen occurs in the barley flower within the spikelets. With this event, pollination occurs followed by fertilization, initiating growth of the embryonic seed or kernel. Although most barley is self-pollinated, natural cross-pollination can occur. In megasporogenesis and microsporogenesis, female and male gametes are developed prior to this event. The mature embryo sac produced by megasporogenesis contains one egg cell (female gamete) and two synergids (cells without walls), and each pollen grain produced by microsporogenesis contains two male nuclei (sperm) and a tube cell nucleus. Pollen is released by rupture of the anthers, falling on the stigma where they germinate. The two male gametes from a pollen grain proceed through the pollen tube into the body of the ovary. One male gamete fuses with two polar nuclei and the egg cell is fertilized by the second male gamete forming the zygote, completing fertilization (Briggs 1978; Reid 1985).

The initial phases of kernel development are complex sequences of events which sometimes occur simultaneously. The result is a progression of cell divisions and the development of specialized tissues forming a fruit (seed) capable of producing a new generation. The barley kernel consists of the lemma and palea, the caryopsis, and the rachilla.

As described previously, the lemma and the palea become the hulls of the mature barley kernel. These structures consists of four types of cells: elongated epidermal cells, blast cells, a thin-walled cell layer of parenchyma tissue, and a thin-walled inner epidermis cell layer. Cellulose, lignin, and silica are the major components of the lemma and palea cells, which are dead at maturity. The hulls are "glued" to the caryopsis, so they adhere to the surface of hulled genotypes but are not attached in hulless types. The cementing substance causing hull adherence in hulled genotypes is secreted by the caryopsis about 16 days after pollination (Harlan 1920). Lemma and palea vary in thickness and in the intensity of which they adhere to the caryopsis, even in hulled genotypes. Thick hulls adhere less firmly to the caryopsis than thin hulls and are less wrinkled after ripening and drying.

The caryopsis consists of the pericarp, the testa (seed coat), the epidermis nucellus, the endosperm, and the embryo or *germ* (Figure 2.6). The pericarp

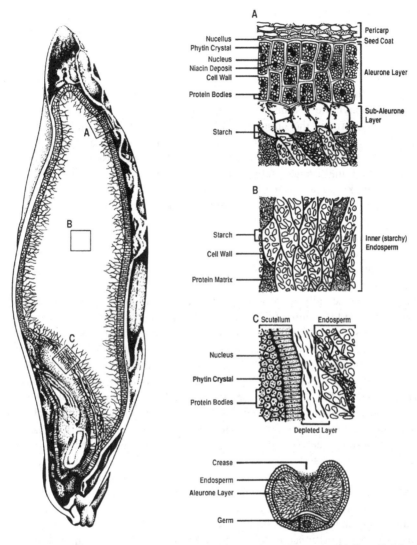

FIGURE 2.6   Barley grain with enlarged cross sections. (Courtesy of Gary Fulcher.)

develops from the ovary wall comprising an outer protective tissue layer, although it is not lignified as are the hulls. It is crushed during development in hulled genotypes, whereas it is not compressed in hulless genotypes. There are usually four types of cells in the pericarp: the epicarp (outer epidermal layer); hypodermic cells; cross cells, which are elongated $90°$ to the kernel axis; and tube cells, which are remnants of the inner epidermis of the ovary wall. The testa lies beneath the pericarp, almost completely covering the kernel, effectively separating the exterior from the interior of the grain. This is a tough membrane composed of cellulose that lies between two cuticularized layers. The outer cuticle layer is thicker than the inner layer. The testa varies in thickness, being thicker over the sides of the furrow and at the apex of the kernel. It is thinner on the sides, over the embryo, and at the micropyle. At the ventral furrow, the testa merges with the pigment strand that runs the length of the kernel, providing a seal at the edges. The remains of the nucellus tissue that surrounded the megaspore and megagametophyte lie just under the testa.

The endosperm is the largest portion of the kernel, consisting of the aleurone, the subaleurone, and starchy endosperm. The aleurone layer is a more intricate structure than the outer integuments. Aleurone tissue is a protein-rich layer of cells found immediately under the nucellus and testa. The cells are separated by thick cell walls and are filled with dense cytoplasm with prominent nuclei, organelles, endoplasmic reticulum, mitochondria, microbodies, proplastids, lipid-containing spherosomes, and complex spherical aleurone grains. Aleurone grains are storage granules containing protein, phytic acid, and hydrolytic enzymes (Briggs 1978). Two large-molecular-weight polysaccharides, arabinoxylan and β-glucan, are structural components of the thick cell walls of the aleurone.

The cellular structure immediately adjacent to the aleurone, designated the subaleurone, is actually more consistent in composition with the starchy endosperm than with the aleurone. The cells of the subaleurone are smaller than those of the starchy endosperm and contain more protein and less starch, which is particularly noticeable in high-protein barleys. The starchy endosperm, making up the largest portion of the kernel, contains cells that are packed with starch grains that are embedded in a protein matrix. Protein is also stored in protein bodies in the protoplasm of the starchy endosperm. The cell walls of the starchy endosperm contain the same two polysaccharides found in the cell walls of the aleurone tissue, but there is a greater amount of β-glucan compared to arabinoxylan.

The embryo, a very complex section of the kernel, is located on the dorsal side of the caryopsis at the end attached to the rachis. The embryo contains the material necessary for initiating the growth of a new plant. It is comprised of the embryonic axis, the plumule, and the radicle, which are surrounded by the coleoptile and coleorhiza, which act as protective sheaths for these tissues. A bud primordial is also located within the coleoptile. The scutellum is a flat protective tissue positioned between the embryo and the endosperm. On the outer side it is recessed so as to fit next to the embryonic axis, and on the inner side it fits against the endosperm. The subcellular constituents of the embryo include mitochondria, protein bodies, spherosomes that contain lipid, Golgi bodies, and rough endoplasmic reticulum, large nuclei, and thin cell walls transversed by plasmodesmata (Briggs 1978). The rachilla, a bristlelike hairy structure lies within the crease on the ventral side of the kernel and is attached at the base where the kernel is attached to the rachis.

The mature barley grain is an elongated oval structure, although more spherical, globular kernels exist among the genotypes. In most cultivated barley, the dorsal side and laterals are rounded and cumulate on the ventral side in a crease sometimes referred to as the *furrow* or *groove* (Figure 2.7). The hull comprises

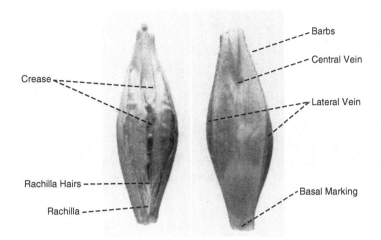

Barbs

Central Vein

Crease

Lateral Vein

Rachilla Hairs

Rachilla

Basal Marking

FIGURE 2.7   Barley kernel characters used for identification. (Courtesy of American Malting Barley Association.)

about 13% of the kernel, the pericarp plus testa about 3%, the aleurone about 5%, the starchy endosperm about 76%, and the embryo plus the scutellum about 3.0% (Briggs 1978; Kent and Evers 1994). Barley kernels vary widely in size due to genotype, position on the spike, and environmental growing conditions. Kernels may weigh anywhere from 5 to 80 mg; however, most hulled barley kernels will weigh between 35 and 45 mg, whereas typical hulless kernels will weigh between 25 and 35 mg. Hulled and hulless kernels are shown in Figure 2.8. The grain is said to be physiologically mature when dry matter accumulation ceases (at about 40% moisture), but the kernel must dehydrate to 13 to 15% moisture to reach harvest ripeness. During the dehydration phase, the pericarp is compressed between the lemma and palea, which adhere to it on the outside and the testa, which adheres to it on the inside in hulled cultivars.

For more detailed descriptions of the anatomy of the barley plant and kernel, the reader is referred to the following publications: Åberg and Wiebe (1946), Wiebe and Reid (1961), Briggs (1978), Reid and Wiebe (1978); Reid (1985); and Kent and Evers (1994).

SUMMARY

Barley evolved over thousands of years from a wild plant to a domesticated (cultivated) cereal grain. A major change from wildness to domestication occurred with the expression of a tough rachis, preventing the seeds from scattering at harvest ripeness. It can be surmised that this event was one of many natural mutations that occurred over thousands of years. The morphological variability seen in cultivated barley is ascribed to changes that occurred during cultivation

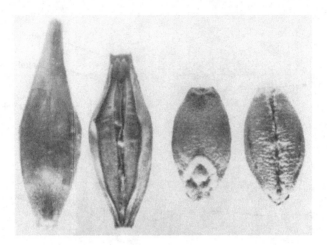

FIGURE 2.8   Hulled and hulless barley kernels. (Courtesy of American Malting Barley Association.)

over an extended period of time in a wide array of geographical areas and to intense breeding. Barley is a grass in the family Poaceae, the tribe Triticeae, and the genus *Hordeum*. The genus *Hordeum* is divided into 32 species totaling 45 taxa. *Hordeum vulgare H. vulgare* L. (cultivated barley) and *Hordeum vulgare H. spontaneum* (wild barley) have identical genomes and are interfertile. The major anatomical parts of the barley plant are typical of those of other members of the Triticeae tribe (wheat and rye), consisting of roots, stem, leaves, spike, spikelets, and kernels. Anatomically, the roots, stems, and leaves of members of the Triticeae family do not differ remarkably among species; however, the barley spike and seeds are very different from those of wheat and rye. The chief identifying taxonomic characteristic of *Hordeum* is its one-flower spikelet, three of which alternate on opposite sides of each node of a flat rachis, forming a triplet of spikelets at each node. There are one central and two lateral spikelets, and all three spikelets may be fertile, or in some, only the central spikelet is fertile. The mature barley grain consists of a hull (lemma and palea), pericarp, testa, aleurone, subaleurone, starchy endosperm, and embryo. The grain is said to be physiologically mature when dry matter accumulation ceases, but the kernel must dehydrate to about 15% moisture to reach harvest ripeness. During the dehydration phase, the pericarp is compressed between the lemma and palea, which adhere to it on the outside, and the testa, which adheres to it on the inside in hulled cultivars. In hulless cultivars, the lemma and palea are in place but do not adhere to the pericarp and testa.

REFERENCES

Åberg, E., and Wiebe, G. A. 1946. Classification of barley varieties grown in the United States and Canada in 1945. *U.S. Dept. Agric. Tech. Bull. 907*.

Åberg, E., and Wiebe, G. A. 1948. Taxonomic value of characters in cultivated barley. *U.S. Dept. Agric. Tech. Bull. 942.*

American Malting Barley Association, Milwaukee, WI 53202. Published online at http://www.ambainc.org.

Bothmer, R. Von, 1992. The wild species of *Hordeum*: relationships and potential use for improvement of cultivated barley. Pages 3–18 in: *Barley: Genetics, Biochemistry, Molecular Biology and Biotechnology.* P. R. Shewry, ed. C.A.B. International, Wallingford, UK.

Bothmer, R. Von, and Jacobsen, N. 1985. Origin, taxonomy, and related species. Pages 19–56 in: *Barley.* Agronomy Monograph 26. D. C. Rasmusson, ed. American Society of Agronomy, Crop Science Socity of America, and Soil Science of America, Madison, WI.

Briggs, D. E. 1978. *Barley.* Chapman & Hall, London.

Fedak, G. 1985. Wide crosses in *Hordeum.* Pages 155–186 in: *Barley.* Agronomy Monograph 26. D. C. Rasmusson, ed. American Society of Agronomy, Crop Science Society of America, and Soil Science Society of America, Madison, WI.

Harlan, H. V. 1920. Daily development of kernels of Hannchen barley from flowering to maturity at Aberdeen, Idaho. *J. Agric. Res.* 19:393–429.

Harlan, J. R. 1978. On the origin of barley. Pages 10–36. in: *Barley: Origin, Botany, Culture, Winter Hardiness, Genetics, Utilization, Pests.* Agriculture Handbook 338. U.S. Department of Agriculture, Washington, DC.

Jörgensen, R. B. 1982. Biochemical and genetical investigations of wild species of barley reflecting relationships within the barley genus (*Hordeum* L.). *Hereditas* 97:322.

Kent, N. L., and Evers, A. D. 1994. *Technology of Cereals,* 4th ed. Elsevier Science, Oxford, UK.

Moral, L. F. G. del, Miralles, D. J., and Slafer, G. A. 2002. Initiation and appearance of vegetative and reproductive structures throughout barley development. Pages 243–268 in: *Barley Science: Recent Advances from Molecular Biology to Agronomy of Yield and Quality.* G. A. Slafer, J. L. Molina-Cano, R. Savin, J. L. Araus, and I. Romagosa, eds. Haworth Press, Binghamton, NY.

Nilan, R. A., and Ullrich, S. E. 1993. Barley: taxonomy, origin, distribution, production, genetics, and breeding. Pages 1–29 in: *Barley: Chemistry and Technology.* A. W. MacGregor and R. S. Bhatty, eds. American Association of Cereal Chemists, St. Paul, MN.

Reid, D. A. 1985. Morphology and anatomy of the barley plant. Pages 73–101 in: *Barley.* Agronomy Monograph 26. D. C. Rasmusson, ed. American Society of Agronomy, Crop Science Society of America, and Soil Science Society of America, Madison, WI.

Reid, D. A., and Wiebe, G. A. 1978. Taxonomy, botany, classification, and world collection. Pages 78–104 in: *Barley: Origin, Botany, Culture, Winter Hardiness, Genetics, Utilization, Pests.* Agriculture Handbook 338. U.S. Department of Agriculture, Washington, DC.

Wiebe, G. A., and Reid, D. A. 1961. Classification of barley varieties grown in the United States and Canada in 1958. *U.S. Dept. Agric. Tech. Bull. 1224.*

3 Barley Biotechnology: Breeding and Transgenics

INTRODUCTION

The question arises as to a proper definition of biotechnology as it applies to plant breeding and transgenics or genetic engineering. In effect, both terms refer to altering the genetic entity of an organism, but through separate and different pathways, to achieve a desired end. Traditional breeding relies on hybridization to transfer genes from one plant to another, relying on phenotypic evaluation to select individual plants with new and more desirable traits. Genetic engineering denotes genetic modification by nonsexual introduction of genes into plant cells via integration into chromosomes using recombinant DNA technology, which can utilize genetic material from the same plant, a different plant, a microorganism, or an animal. With both traditional breeding and recombinant DNA technology, the genetic material in the cell is rearranged. The use of recombinant DNA technologies to modify plants is commonly referred to as *genetic engineering*, and in the popular press the product is often referred to as a *genetically modified organism* (GMO).

The purpose of barley breeding or genetic engineering is to produce an improved version of a barley plant by introducing and expressing one or more desirable genetic characteristics from a donor plant or organism to a receptor plant. According to Anderson and Reinbergs (1985), barley breeding as the science we know today began early in the twentieth century. It has been suggested that a proper choice of parent barleys and progeny is not a defined science, but rather, that success often depends on the perceptive intuition of the breeder. Early efforts in breeding were directed almost entirely at selecting barleys with characteristics that increased the total yield of grain harvested. Yield continues to be an important objective of selection, as high yields are obviously essential for economic reasons. However, early on, scientists began to recognize that the expression of certain genetic traits in high-yielding elite cultivars was desirable

Barley for Food and Health: Science, Technology, and Products,
By Rosemary K. Newman and C. Walter Newman
Copyright © 2008 John Wiley & Sons, Inc.

or even essential in the end product, the barley kernel. Hence, barley breeders began to include emphasis of such characteristics in selection programs as they were identified. This has become especially important for malting quality traits, such as enzymatic levels and various aspects of the kernel composition, as the utility of barley for beverage production has become more valuable than the utility of barley for feed. Modern barley cultivars used for malting and brewing are selected as much for their malting characteristics as they are for their productivity characteristics. The two criteria are often at odds with one another, requiring close attention of the breeder.

According to Wiebe (1978), "progress in barley improvement depends on a supply of good genes and on breeding techniques for assembling these into superior genotypes." Until the concept and successful advent of molecular genetics and nonsexual gene transfer in barley by three separate research laboratories in 1994, genetic improvement of barley was limited to traditional plant breeding practices. The primary sources of genotypic variation for barley breeding prior to 1994 included existing cultivars, breeding lines, mutations, and exotic gene sources. The latter consisted of landraces of cultivated barley (*Hordeum vulgare* subsp. *vulgare*) and genotypes within the related subspecies *H. vulgare* subsp. *spontaneum*. *H. bulbosum* represented a secondary gene pool source but has limited use, due to sterility in the hybrids produced (Bothmer et al. 1995). The primary gene pool has been almost the sole source of allelic variation for producing commercial cultivars with limited inputs from the secondary gene pool. Enhancement of commercial barley germplasm to date has been limited to crossing elite breeding lines and cultivars and will probably continue to be the primary method of genotype improvement for at least the immediate future (Lemaux et al. 1999). However, the increased knowledge and technique improvements in mutagenic breeding and transgenic cereal research exemplified by Wan and Lemaux (1994) offer exciting opportunities to bypass many of the limitations of traditional breeding practices and to provide access to more diverse sources of genes.

IMPROVING THE BARLEY CROP

Selection of barley plants having one or more desired characteristic(s) no doubt occurred for eons prior to the science and knowledge of genes and heritability. It can be assumed that in the beginning stages of agriculture it was recognized that "like begets like," resulting in the development of individual barley landraces typical of a given area. It is likely that landraces developed in various regions of the land were rarely, if ever, mixed until trade routes evolved between local settlements and distant sections of the world. Throughout most of the history of barley development and utilization, food and beverage production were the important end uses. Therefore, those barleys that were palatable and/or were successful in making beverages were probably those grains that were propagated. As barley foods were gradually replaced by wheat and other grains, barley's primary use was shifted to animal feed, along with continued use in beverage production. As mentioned previously, breeding objectives in barley improvement have been strongly influenced by the malting and brewing industries rather than the feed market, as malt-quality barley demanded a premium price in the market. In the past 50 or so years, most barley breeding has been subsidized by the malting and brewing industries, a trend that continues. It is only in relatively recent times that any emphasis has been placed on feed quality breeding (Ullrich 2002), although selection of plump kernels with a high starch content, a characteristic for malting quality, is also beneficial to feed quality. Even more recently, scientific demonstrations of the health, medicinal, and nutritional aspects of barley, most notably, but not limited to, the high levels of β-glucans in the barley kernel, have stimulated considerable interest in utilizing barley germplasm for new and improved food products.

Improving the genetic profile of barley basically involves assembling the myriad versions, or alleles, of genes that interact in complex ways to produce optimal combinations of desired quality traits in plants that are acceptable for agronomic production in a given environment. The breeder must first define the goals of the breeding program, weighing the value of each trait or traits desired against the cost in time and money required to achieve the objectives. After goals are defined, an appropriate breeding program may be selected that will best meet the required purpose(s) in producing a cultivar in an economically competitive genotype. Economic competitiveness is a combination of gross productivity and the economic value of the quality traits that is realized. Malting or food cultivars, for instance, can be competitive economically at lower levels of productivity than can cultivars used strictly for feed.

Traditional Barley Breeding

Methods and advancements in traditional methods of barley breeding have been presented in numerous scientific essays, book chapters, and professional journals. The reader is directed to the following publications for detailed history and techniques in barley breeding technology (Harlan and Martini 1938; Briggs 1978; Wiebe 1978; Starling 1980; Anderson and Reinbergs 1985; Swanston and Ellis 2002; Thomas 2002; Ullrich 2002). These publications are a sampling of the literature available on this subject, much of which can be found on the Internet. Traditional approaches to breeding for improved barley production and quality for the past 100 years include conventional breeding, mutagenic procedures, haploid production, interspecific and intergeneric crosses, and molecular marker-assisted selection breeding.

Conventional Barley Breeding

In breeding barley, scientists are faced with deciding on the choice of parent barleys and later, with choosing the resulting segregates to save for future generations. In addition to determining the value or effect of a particular allele, the scientists must select the most appropriate available breeding method. Plant breeding is not unlike animal breeding in many respects; there is an old saying that "the eye of the master fattens the cattle," which in essence describes the barley breeder's intuition for choosing the genetic backgrounds and appropriate breeding methods best suited to select and assemble the best genes in a new plant.

Crossing cultivars of different genetic backgrounds is perhaps the most basic method in barley breeding. As barley is a self-fertilizing plant, artificial crosses are required to produce recombinant plants (Wiebe 1978). Controlled crossing requires basic knowledge of plant morphology and the ability to recognize the progression of events from early floral development through pollination, a degree of physical skill necessary to dissect tender plant tissues without destroying them, basic knowledge of maturity characteristics of the parents that are to be hybridized, and patience. One difficulty that has to be overcome in many instances is the difference in pollen-maturity dates (days to anthesis or pollen shedding) that may exist between proposed parents. As barley is a self-pollinating plant, emasculation of the female parent is required either surgically, which is time consuming, or by using male sterile plants. A number of genes producing male sterility in barley have been documented (Hockett and Eslick 1968). Cross-pollination can be accomplished by a number of methods, some requiring bagging the female spike to prevent accidental pollination by foreign pollen.

It is generally agreed by breeders that crossing barley plants that are known to be high yielding will produce high-yielding crosses with low genetic variance in most instances. Intercrossing plants with a restricted range of parental lines can reduce the number of gene pairs segregated, thus preserving previous genetic advances while providing a reasonable chance of improving specific traits (Eslick and Hockett 1979). The disadvantage of this approach to improving various traits is that it leads to a restricted gene pool (Anderson and Reinbergs 1985). Despite such theoretical and demonstrated losses in genetic diversity that are the consequence of limited parental selection, decades of selection and restriction have nevertheless not prevented continued gains from selection (Rasmusson and Phillips 1997; Condon et al. 2008).

Backcrossing, a technique first outlined by Harlan and Pope (1922), has been used extensively over the years in many breeding programs. It is well suited for transferring simply inherited characters controlled by one or two major genes (Wiebe 1978) such as the *nud* (hulless) gene. Backcrossing involves repeated backcrossing to one recurrent parent after an initial cross to another (donor) parent containing the simply inherited (one- or two-gene) trait of interest, the aim being to recover only the donor parent trait in a genome that becomes increasingly similar to the recurrent parent with each successive backcross. Taken to the extreme, backcrossing creates a nearly isogenic line, identical to the recurrent parent in all aspects except for the desired donor-parent characteristic. In practical terms, the number of backcrosses that can be performed is limited by time and resources, and it is generally recognized that isogenic pairs differ by small gene blocks rather than a single gene (Wiebe 1978) and lines are sometimes referred to as *isotypes* rather than isogenic pairs. Backcrossing is widely employed for the introgression of disease resistance, for instance, and for isogenic analysis: that is, the comparison of two alternate states (alleles) of a gene in an otherwise homogeneous background in the same or nearly the same background genotype. For example, isogenic lines contrasting six- and two-rowed spikes have been developed and compared for agronomic and malt quality (Wiebe 1978). Similarly, hulled and hulless isolines have been developed in several genotype backgrounds to measure the influence of dietary fiber on the feed value of barley (Hockett 1981).

Bulk handling is an efficient and inexpensive method to grow large numbers of early cross-line generations. Selections are made and tested as individual lines at the end of the scheduled breeding program. The bulk breeding system is adapted to mass selection and is useful in identifying numerous phenotypic characteristics. Composite crossing is a type of bulk breeding where a number of single crosses are combined into a composite mixture, providing an efficient selection method. When using the composite crossing technique, projected objectives are generally long term in nature, allowing for recombination of many genes from a broad-based germplasm (Anderson and Reinbergs 1985). A breeding technique termed *male-sterile-facilitated recurrent selection* (MSFRS) was developed by R. F. Eslick at the Montana State University and Agricultural Experiment Station. When used in conjunction with the composite crossing technique, MSFRS allows selected matings within composites to achieve specific breeding objectives (Ramage 1975). A deviation of this method is the *diallel selective mating system* (DSMS) proposed by Jensen (1970). In this system, simultaneous insertion of multiple genotypes can be introduced into a few controlled populations. MSFRS and DSMS facilitate broadening of the genetic base and break up existing linkage blocks while providing greater genetic variability than is available in many other breeding systems (Anderson and Reinbergs 1985).

Several widely used methods that may be considered as conventional barley breeding programs include single-seed descent breeding (SSD) and pedigree breeding (Tourte 2005). The *SSD breeding method* was proposed as a way to maintain maximum genetic variation in self-pollinating species while obtaining a high level of homozygosity. This method may be used for parental evaluation, which is accomplished by evaluating an array of homozygous lines from several crosses and identifying those crosses that have the highest proportion of superior progeny. *Pedigree breeding* is best applied where genetic characters are highly heritable and can be identified in early segregating populations but not for characters with low heritability. Pedigree breeding is the most common method employed for characters with complex inheritance, such as malting quality. Breeders working within narrow germplasm pools regularly use pedigree breeding–based methods, frequently in combination with the SSD method to get many lines into advanced states of homozygosity without losing genetic variability. A strong point of the pedigree system is the ease with which the planned breeding can be modified at any stage of selection (Lupton and Whitehouse 1957). Modification of the pedigree system as proposed by these authors provides for yield estimates at the same time as line generations are being advanced.

Mutations

When used in connection with plant breeding, the term *mutant* or *mutation* has a hazardous and even an intimidating connotation that often causes unwarranted concern. Mutant barley plants occur spontaneously (naturally) as a result of errors in normal cell processes (e.g., DNA replication), may be induced by a number of chemical or radiation treatments, or may occur following cell and tissue culture propagation. Mutagenesis research has been reported and reviewed extensively in the scientific literature (Hockett and Nilan 1985). Regardless of the mutations source, mutant barleys should not be considered as dangerous or undesirable. In effect, natural and induced mutations have been a mainstay in producing genetic variation since the advent of green plants; without genetic variation, improvement of a species would not be possible.

Mutant barley may be utilized directly, providing an "instant" cultivar, but more often, potentially useful mutations (especially when induced via chemicals or radiation) are accompanied by other mutations, which must be removed by an appropriate breeding method such as backcrossing. Both spontaneous and mutant variants may provide evidence of genes at previously unknown loci as well as new alleles in improved adapted genetic backgrounds, thus supplementing and complementing a conventional breeding program. Mutagenesis has been used extensively for development of new and improved cultivars and has provided new methods of analyzing genetic fine structure and understanding genetic control (Hockett and Nilan 1985). A relatively recent development in mutation strategy enables a directed approach to identifying mutations in specific genes. This procedure is described as *targeting induced local lesions in genomes* (TILLING). Using this strategy, single-base-pair changes can be identified in a specific gene of interest (McCallum et al. 2000; Till et al. 2003). Slade et al. (2005) identified a range of waxy phenotypes in the wheat TILLING population, suggesting that similar starch variants may be identified in barley. Using the TILLING technique, point mutations and deletions in the powdery mildew resistant genes in barley, *mlo* and *mla*, have been identified (Mejlhede et al. 2006), presenting an opportunity to study resistance to these parasitic fungi.

A mutant barley resistant to powdery mildew was produced over 60 years ago by Freisleben and Lein (1942) (cited by Ganeshan et al. 2008). As of 2007 there were 256 various mutant barley varieties listed in the FAO/IAEA Mutant Varieties database (FAO 2007). A number of mutant barleys have been identified with genotypes expressing useful genetic characters—with important implications for the food and brewing industries. The genetic background in the spontaneous high-amylose starch mutant found in Glacier by Merritt (1967) may prove to be invaluable in developing high-amylose starch barleys for use in specialty foods. Induced mutants are also sources of valuable traits, such as increased kernel protein (Doll 1973), high lysine (Jensen 1979), proanthocyanidin-free (Jende-Strid 1976), high β-glucan (Eslick 1981), low β-glucan (Aastrup et al. 1983), and high-amylose starch (Merritt 1967; Morrell et al. 2003). The high-β-glucan barley (Prowashonupana) produced by Eslick (1981) is an example of an instant cultivar, as it is now being grown commercially without further modification.

This barley is being utilized as a source of soluble dietary fiber for specialty foods. The lysine gene *lys3a* from the Danish high-lysine mutant Bomi 1508 (Jensen 1979) was transferred successfully into a high-yielding, high-lysine feed cultivar through a series of cross-breeding programs. Feeding trials with mono-gastric animals have effectively demonstrated the nutritional quality of the protein in this barley (Munck 1992). Identity preservation is essential when growing spe-cialty cultivars, which may add significantly to the cost of production. The cost must be weighed against the value of the product. In areas of the world where quality protein is in short supply, high-lysine barley would be invaluable as a source of protein for humans.

Haploids

Haploid plants have one set of complete chromosomes representing the basic genetic complement, which is one-half of the normally observed diploid chro-mosome number in an organism. Methods of production and uses of haploids in barley breeding have been reviewed (Kasha and Reinbergs 1976; Anderson and Reinbergs 1985; Fedak 1985; Pickering and Devaux 1992). Methods of producing haploids include ovary culture (in vitro gynogenesis), chromosome elimination (*bulbosum* technique), anther culture (in vitro androgenesis), microspore culture, interspecific hybridization, and use of a haploid initiator gene (*hap*) that was found by Hagberg and Hagberg (1980).

Production of double haploids from the original haploid is necessary for the system to work. In the case of haploids-derived anther cultures, spontaneous doubling often occurs (Luckett and Darvey 1992). In other systems, where spon-taneous doubling occurs rarely, treatment with colchicine, a toxin derived from autumn crocus, *Colchicum*, efficiently produces double haploids, allowing for the recovery of homozygous inbred lines in a single generation. Pickering and Devaux (1992) suggested several applications of haploids in addition to producing new cultivars: early identification of superior hybrid combinations, detection of linkages associated with quantitatively inherited characters, calculation of recom-bination values between linked genes, and evaluation of pleiotropic (multiple) effects of specified genes.

In comparing haploid breeding with single-seed descent and pedigree breeding it was concluded that growth performance means, ranges, and frequencies of superior genotypes were similar (Park et al. 1976; Reinbergs et al. 1976). Each method was superior in one or more comparative measures; however, when the 10 best lines from each method were compared, there were no differences in yield (Kasha and Reinbergs 1979). In a comparison of seven crosses using pedigree and doubled-haploid derived lines, Turcotte et al. (1980) reported that superior lines were produced by the haploid technique.

The culture of haploids is a growing field in the plant improvement sector, having the advantage of producing plants that are totally homozygous and con-stituting pure lines that can be used in pedigree or other methods of breeding. Additionally, ease of selecting among homozygous lines and time savings are recognized advantages of the haploid technique. The size of breeding populations can be reduced while preserving a probability of success similar to that of other, more labor-intensive and time-consuming methods. In practice, these techniques have been used to select characters in barley related to yield, size, parasite resistance, and early maturity. A significant number of barley cultivars have been released as double haploids (Tourte 2005).

Interspecific and Intergeneric Crosses

Various approaches to interspecific and intergeneric breeding of barley have been investigated by a number of researchers for the purpose of introduction of new alleles into barley. This topic has been the subject of several reviews (Fedak 1985; Bothmer 1992; Shepherd and Islam 1992; Ellis 2002).

With the exception of *H. bulbosum*, species other than *H. vulgare* show little or no chromosome pairing at meiosis, and little or no opportunity for interspecific gene transfer exists. Incompatibility traits of *H. vulgare* crosses with other species of *Hordeum* were detailed in a review by Pickering and Johnston (2005). The major incompatibilities pointed out were chromosome elimination, endosperm degeneration, infertile hybrids, reduced recombination, linkage drag, instability of hybrids, and chromosome pairing. Because *H. vulgare* and *H. bulbosum* chromosomes show pairing at meiosis (Bothmer 1992), interspecific gene transfer is possible but has proven to be difficult. However, powdery mildew resistance has been transferred to *H. vulgare* from *H. bulbosum* (Pickering et al. 1995). Crosses between these two species have been widely employed as a means to produce haploid *H. vulgare* plants since Kasha and Kao (1970) described the elimination of *H. bulbosum* chromosomes in interspecific crosses.

H. spontaneum, an annual two-rowed diploid, is the only wild *Hordeum* species that is totally compatible in crosses with cultivated barley and is considered a subspecies of *H. vulgare*. Being subspecies, mating of *H. spontaneum* with *H. vulgare* is not truly an interspecific cross, however, it is included in this discussion since it is a source of genetic variation for use in barley breeding programs. The primary morphological difference between *H. vulgare* and *H. spontaneum* is that the ears of the latter have brittle rachis and at maturity disarticulate into arrowlike triplets designed for self-planting, ensuring a new crop. Crosses of *H. vulgare* and *H. spontaneum* are completely interfertile and show normal chromosome pairing and segregation at meiosis. Spontaneous hybridization occurs sporadically between *H. vulgare* and *H. spontaneum* when grown together, producing completely fertile hybrids and demonstrating the genetic affinity between the two barley forms (Bothmer 1992; Nevo 1992).

Intergeneric hybridization (i.e., barley × wheat crosses) has provided additional sources of genetic material. Fedak (1985) presented a review of research accomplished and published on intergeneric breeding of cultivated barley and relatives within the Triticeae tribe. More recently, comprehensive reviews have been published on this subject (Islam and Shepherd 1990; Fedak 1992; Shepherd and Islam 1992). The phytogenetic relationship between genera within the tribe Triticeae was the stimulus to geneticists to obtain hybrids between barley and wheat. Little success was achieved in this area until 1973, when it was demonstrated that diploid *H. vulgare* L. could be crossed as a maternal parent with diploid, tetraploid, and hexaploid wheat (Kruse 1973). Numerous combinations of barley × wheat crosses have since been developed (Fedak 1992). Wheat–barley recombinant lines are projected to provide useful supplementary information for mapping of the barley genome, including further understanding of genome organization and evolution in these two cereals. Interest in barley–wheat hybrids is also directed toward translocation of specific genetic traits into barley from wheat, or reciprocally, wheat traits into barley, providing the potential of creating a new cereal such as triticale, which was produced by crossing durum wheat with rye.

Molecular Markers

Molecular marker–assisted selection (MAS) is a laboratory approach to barley breeding whereby chromosomal regions (loci) controlling particular traits of interest can be identified and followed in a selection program. Effective use of this technique in breeding demands the construction of chromosome linkage maps and association of unique biochemical markers that are associated with different alleles at these loci. In some cases, loci associated with traits can be associated with a particular gene or genes, but the method can be used effectively, even with no information about the gene or genes, as long as they are linked genetically to the biochemical marker used for selection. There are significant limitations to MAS for the selection of complex quantitative traits. Interactions between large numbers of potential alleles at a large number of loci controlling a trait make MAS more suitable for single traits or those controlled by only a few genes. Certain aspects of disease resistance and quality are good candidates for MAS, especially those that are difficult or expensive to phenotype (P. Bregitzer, personal communication).

Molecular mapping of the barley genome began in the 1980s (Kleinhofs et al. 1988), and since that time, extensive data have been generated by numerous researchers that have permitted the development of detailed barley chromosome linkage maps. The biotechnology leading to the mapping of the barley genome was facilitated by the development of molecular markers, double haploids, availability of numerous mutants, intergeneric crosses, and insert libraries (Kleinhofs and Han 2002). The North American Barley Genome Project (NABGP) is a continuing, multi-institutional, multidisciplinary project that has developed extensive germplasm and molecular resources for genetic mapping, gene isolation, and gene characterization in barley. Core mapping populations and specialized germplasm resources allow NABGP researchers to place thousands of markers on genetic maps (CRIS USDA 2007). The Barley Coordinated Agricultural Project (Barley CAP 2007) is also a group effort currently composed of 30 or more barley researchers from 19 institutions with expertise in genetics, genomic, breeding, pathology, databases, computer science, food science, malt quality, and statistics.

The overall theme of the barley CAP is to integrate and utilize state-of-the art genomic tools and approaches in barley breeding programs to facilitate the development of superior barley cultivars. Recent efforts have also focused on the development of functional genomic tools using the maize transposable element, *Ds*, to perform targeted mutagenesis and reactivation tagging of genes in barley (Cooper et al. 2004; Singh et al. 2006).

Traditional and conventional breeding of barley has not been replaced by the use of genetic maps and marker techniques, but rather, these newer approaches add to the tools that breeders can utilize in the identification and location of genes and their effects on trait expression (Rao et al. 2007). Simply said, recent developments in molecular techniques for barley breeding does not replace tried and true breeding approaches, but instead, provides methods for conformation and extension of traditional breeding programs. The most comprehensive updates in current developments in mapping strategies and associated genetic research on the barley genome may be found in annual online publications of *Barley Genetics Newsletter* (www.wheat.pw.usda.gov/ggpages/bgn).

BARLEY TRANSGENICS

Barley was a difficult cereal grain upon which to apply transgenic technology successfully, but not more so than wheat. Major obstacles have now been overcome using direct and indirect methods of transformation. Among the numerous pioneers who produced transgenic barleys, Wan and Lemaux were first to report efficient transformation of barley that resulted in fertile, stably transformed plants (Wan and Lemaux 1994). It is also notable that two other research groups using different approaches reported similar successes in the same year (Jähne et al. 1994; Ritala et al. 1994). With this success, barley was one of the last of the major cereal grains in which fertile transgenic plants were produced successfully. The opportunities that this accomplishment provides for the future of barley breeding should neither be ignored nor wasted. Bringing laboratory successes in transgenic research to the supermarket and dinner table will not be an easy accomplishment given the vastness of misconceptions to overcome. However, with continued dedication of researchers in genetic transformation and a positive attitude of barley believers, this method should become an accepted tool in the arsenal of the barley breeder. At least five extensive reviews have been published on the history, methodology, and progress of transgenic barley research citing the difficulties encountered and successes in introducing foreign genes into fertile, stably transformed barley (Lemaux et al. 1999; Jacobsen et al. 2000; Horvath et al. 2002; Jansson and Ferstad 2002; Ganeshan et al. 2008).

Why Transgenics?

Transgenics is a term describing plants that are manipulated using a genetic modification technique, also called genetic engineering, that introduces genes from any organism: from barley itself to other cereals or plants, bacteria, and even humans. Why transgenic barley? Classical or traditional plant breeding can do just so much with currently available alleles in compatible germplasm in terms of continued improvements in barley crop yields, while enhancing malt quality and nutritional value for food and feed. The application of genetic engineering to cereal grain breeding has the potential to expedite the solution to many problems facing the agricultural and food industries, such as conserving natural resources and protecting the environment while providing abundant economical products to the consumer. Considering the current worldwide population explosion, it is imperative that genetic engineering of barley and other cereals be promoted and extended. Collectively, cereal grains account for 66% of the world food supply (Borlaug 1998), and according to Vasil (1999), food production must be doubled from the turn of the century to 2025 and nearly tripled by 2050 to meet the food needs of the twenty-first century. It is doubtful that such increases in world food production can be accomplished in such a short period of time by traditional breeding alone. For example, the development of a new and improved variety of a cereal grain may take 10 or more years using traditional breeding practices. Today, that time span can be reduced significantly by using the best available tools. Careful selection and introduction of desirable genes from other plants or organisms into the barley genome provides the necessary avenue for continued improvement of this important crop.

Yield is the single most important criterion for a cereal crop, including barley, but yield in itself is not due to a single attribute. Improved yield is controlled by many factors, resulting from improved crop management strategies in addition to plant selection for superior genotypes that are resistant to diseases, pests, lodging, shattering, drought stress, and other environmental hazards. Enhancing these traits in barley is within the reach and abilities of scientists trained in the science and techniques of genetic engineering and breeding. In addition to producing better barley plants, genetic engineering provides a new way to produce nutritionally improved foods for areas of the world where barley is a mainstay for food and feed. Genetic engineering can be used to create specially designed barley foods to provide for the dietary needs of individuals susceptible to food allergies or who suffer from diabetes, cardiovascular disease, or other food-related illnesses (see Chapter 8). Specially designed hulless barleys, which provide high-protein nutritionally enhanced feedstocks, can be engineered to increase biofuel production. Through genetic engineering, there is the potential, safely and economically, to produce many useful compounds, such as enzymes, antibiotics, hormones, and other biological agents, utilizing completely environmentally friendly methods.

Delivery Systems

Methods of transforming plant cells may be classed as either indirect or direct. *Indirect transformation* involves the use of a soil phytopathogen, *Agrobacterium tumefaciens*, that induces tumors known as crown galls, primarily on dicotyledonous plants. There are several known strains of agrobacteria; however,

A. tumefaciens is presently the most studied and best known of these bacteria. *A. tumefaciens* is the only known prokaryotic organism that can transfer DNA to eukaryotic cells, and this remarkable capability has been exploited as a method of controlled transfer of genes from one plant to another. *A. tumefaciens* has been used routinely since the first published success in tobacco and tomatoes (Horsch et al. 1985) of transferring genes to dicotyledonous plants, but similar successes with monocotyledon plants were not achieved until it was discovered that inducing *Agrobacterium*-mediated transfer in monocots required different approaches. Development of optimized protocols included several critical factors, including specific *Agrobacterium* strains and vectors, coupled with appropriate selectable marker genes, selection agents, plant genotypes, and type and age of explants (Komari and Kubo 1999). Within the last 10 years, the use of *Agrobacterium* for transformation has become a reality for barley (Tingay et al. 1997; Guo et al. 1998; Wu et al. 1998).

Direct transformation refers to the movement of genes from one organism to another without an intermediary organism such as *Agrobacterium*. Injection and incorporation of isolated DNA from a normally growing barley plant into a second similar barley plant to effect transfer of genetic material would be an ideal transformation system that would eliminate in vitro culture and the associated problems of regeneration and somaclonal variation. Unfortunately, several attempts using variations of this approach were unsuccessful (Ledoux and Huart 1969; Soyfer 1980; Töpfer et al. 1989; Brettschneider et al. 1990; Mendel et al. 1990). Transfer of foreign DNA into barley has been achieved to some degree of success by microinjection and electroporation procedures as reviewed by Karp and Lazzeri (1992) and Ganeshan et al. (2008).

Microinjection is a procedure whereby foreign DNA is injected into a cell or protoplast using a micropipette mounted on a micromanipulator. Microinjection requires specialized skills, expensive equipment, and resources. In *electroporation*, cell walls are removed and protoplasts are transformed by adding them to a concentrated solution of plasmidic DNA with the desired insert in a chamber with electrodes, between which an electric field is applied for a few milliseconds to create holes through which the DNA can enter. The main problem with this procedure is regenerating plants from protoplast, as sometimes the cells "forget" that they are plant cells (P. G. Lemaux, personal communication).

Microprojectile bombardment of explants (plant tissues) with DNA-coated particles which are subsequently regenerated during in vitro tissue culture is currently the method of choice for producing transgenic barley (Tourte 2005; Ganeshan et al. 2008). This method of transformation is rightfully termed *biolistics* since the primary mechanism utilized, which was conceived in 1984 by John Sanford, a professor at Cornell University, is described as a "gene gun." The evolution of this instrument, research applications, and the techniques involved in its use to transform cereals, including barley, are discussed in detail by Klein and Jones (1999). In the original studies, Sanford used a 22-caliber rifle with gun powder–charged cartridges to propel tungsten microbullets coated with DNA into a plant leaf target. Since the conception of this idea, the instrument has been refined considerably and the method has evolved into an important approach for creating transgenics in grasses (Gramineae/Poaceae). The basic function of the instrument is to propel DNA-coated particles positioned on microcarrier material (a Mylar disk) through a tube (barrel) via a shock wave created by a burst of compressed helium. The microcarrier is propelled through a partial vacuum and strikes a stopping plate, but the DNA-coated particles continue down the barrel and strike the target tissue, where the DNA comes off and in some cases integrates into the plant genome. This is a very brief and nonscientific description of a very complicated and sophisticated instrument and procedure. The success of the outcome is affected by numerous critical variables, such as microprojectile particle size, the number of particles that strike the target tissue, the momentum of the particle, and the in vitro culture system used.

Target Tissues

Barley protoplasts were the initial choice for transformation targets using electroporation; however, there were many problems associated with their use. Therefore, other target tissues were explored, including apical meristems, cultured shoot meristematic tissue (Zhang et al. 1999), callus from young leaf tissue, green highly regenerable, meristemlike tissue (Cho et al. 1998), immature scutella (cotyledon tissue), immature ovaries, immature inflorescences, and anthers/ microspores as reviewed by Tourte (2005) and Ganeshan et al. (2008). Gürel and Gözurkirmizi (2002) used mature barley embryos with electroporation, producing stable transformed plants. Ganeshan et al. (2003) used microprojectile bombardment of mature barley embryos to produce multiple shoots directly, which were shown to be transformation competent. These authors credited success in genetic transformation with mature embryos in part to the development of specialized media in tissue culture. In an earlier report, Zhang et al. (1999) found many of the same advantages in genetic transformation using mature seeds to yield cultured meristematic tissue. A high frequency of regeneration was observed in mature embryos in this study and reported to be "fairly genotype independent." Several advantages were listed for use of mature embryos: There is no requirement for the growth and maintenance of donor plants, mature embryos can be readily isolated from mature seeds, and vernalization (exposure to cold temperature) is not required in the case of winter barley. From the viewpoint of these researchers, the mature embryos presented a simple, efficient, and easily obtainable culture system for transformation of barley plants.

In Vitro Culture

Success in transformation procedures in most cases depends largely on utilization of the proper in vitro tissue culture methods that result in regeneration of genetically engineered plants. Development of in vitro tissue culture methodology began at the beginning of the twentieth century, being at that time more of a visionary idea to understand a cell's ability to regenerate a complete plant. The impact of the in vitro plant tissue culture research of the past century was fundamentally important to rapid development in the genetic engineering of plants. Novel tissue culture systems were developed to counter many of the problems that arose in the development of transformation methods for commercial barley cultivars.

Tissue culture systems used for transgenics consist of two aspects, the explant (plant tissue) treated with foreign DNA and the culture medium to which the explant is exposed. Procedures call for sterilization of the medium to prevent bacterial, viral, and/or fungal contamination. Culture media vary in nutrient and hormone composition, depending on the nature of the explant and type of tissue desired. In general, media contain macro- and micromineral salts in various concentrations, amino acids, and vitamins. Phytohormones may be added to the base medium in variable proportions. Explants are cultured on the medium in test tubes or petri dishes and at particular points are exposed to low light illumination. Other parameters, such as temperature and humidity, are also controlled as much as possible (Tourte 2005).

Altered Traits

Transgenic studies on barley have increased the overall understanding and knowledge of barley tremendously. The use of molecular and genomic tools, including genetic engineering, has played an important role in analysis of the genome and the development of an understanding of its critical functions with regard to the growth and development of plants. Although there have been immense strides forward in using transgenic technology on barley and other cereals, the science is truly in its infancy, considering its potential. Genetic engineering, implemented correctly, can change the face of agriculture in many ways beyond the simple concept of crop yield. Over 40% of crop (cereal grain) productivity is lost to competition with weeds and to pest and pathogenic diseases, plus an additional 10 to 30% in postharvest loss. A major percentage of these losses can be eliminated or reduced substantially by using transgenic biotechnology (Oerke et al. 1994; Vasil 1999).

Ganeshan et al. (2008) tabulated 23 reports describing attempts at trait modification in barley by genetic engineering. Of these reports, 10 dealt with increasing disease resistance, six with improving malting quality, four with nutritional quality, and three with other factors in crop production. Malting quality efforts were primarily designed to express thermostable β-glucanase (Jensen et al. 1996), increase the activity of thermostable β-glucanase (Nuutila et al. 1999; Horvath et al. 2000), increase thermostable β-amylase activity (Kihara et al. 2000), increase α-amylase activity (Tull et al. (2003), and speed α-amylase and increase pullulanase activity (Cho et al. 1999). Zhang et al. (2003) were not successful in changing grain hardness by inserting the zein gene. The genotypes used in these studies were Golden Promise, Kymppi, and Ingri. Four reports on nutritional quality alterations in barley were aimed at increasing free lysine and methionine content (Brinch-Pedersen et al. 1996), xylanase activity (Patel et al. 2000), cellulase production (Xue et al. 2003), and short-term zinc uptake (Ramesh et al. 2004). The efforts to increase β-glucanase expression and activity have major implications on improving the nutritional value of barley for poultry feeds by negating the necessity of adding exogenous β-glucanase to the feed. From a human nutritional standpoint, this has negative and positive connotations, as will be discussed in Chapter 8.

In addition to the genetic engineering efforts listed above, 10 barley genes were isolated by map-based cloning and at least 18 barley-specific patents were applied for from 1997 through 2006 (Ganeshan et al. 2008). Protecting intellectual property with patents has received considerable attention in recent years. It is therefore not surprising that numerous patents, both issued and proposed, deal with techniques for transformation of barley.

Transgenic Barley Risk: True or Imagined?

There are risks with regard to the safety of food products regardless of the method used in the development, whether it is by one or more traditional methods or by modern genetic engineering technology. A question often posed in opposition to the use of transgenic technology in plants utilized for human food involves the potential for movement of a transgene from the food into the human genome following consumption of the food. The possibility of this occurring is quite remote, nearly impossible in normal individuals. When the DNA and messenger ribonucleic acid (RNA) of the target gene are ingested, they are broken down (digested) to their basic components: purine or primidine bases (thymine, cytosine, adenine, and guanine), a pentose sugar (ribose), and phosphoric acid. Thus, the genetic material is no longer capable of functioning as a gene. Insertion of marker genes that sometimes encode antibiotic resistance are used to identify the cells containing the target gene. The use of marker genes designed in this manner has received criticism due to the possibility of transferring antibiotic resistance to microflora in the human gut. Aside from being subjected to enzymatic digestion, eukaryotic cells are naturally resistant to most antibiotics that are targeted to bacteria, and most antibiotics used in plant transformation are not currently used in medicine (Tourte 2005). Resistance of gut bacteria to antibiotics has occurred, but it is more likely to be due to overuse of antibiotics in clinical situations and in animal feed.

Behavior of a transgene when introduced into a new environment cannot be predicted accurately. The behavior of a transgene may vary, resulting in amplification, cosuppression, or disruption of a functional gene (Tourte 2005). Although possible, the likelihood of activating toxins or creating allergens in an engineered variety is no more likely than its occurrence in a plant created by classical breeding. Despite the low likelihood, all engineered crops in commercial production are tested exhaustively for substantial equivalence, meaning that they are identical to the parent crop, with the exception of the trait introduced. If a gene is introduced into a crop from a known allergenic source such as peanuts or wheat, FDA rules state that the food must be labeled as containing the gene from that variety. Does this mean that genetically engineered foods have zero risks? No, and this cannot be said of foods created by classical breeding as well. This is exemplified by certain potato and celery varieties created by classical breeding that were found to contain human toxins; fortunately, because of laboratory testing, they never reached the consumer market.

The answers to these and other consumer concerns require testing and observation of the engineered plant. Thus, consumers are faced with considering the risk–benefit of utilizing and consuming the products of this new technology, realizing that neither risks nor benefits can categorically be assumed to be absent. In assessing risks and benefits, for some consumer activists, the goal is to achieve *zero* risk, which although desirable is unfortunately unattainable in any biological system. Most food scientists agree that small risks are inherent in our abundant food supply, which yields ample and inexpensive nutrition to everyone (Jones 1992). From a nutritional health benefit viewpoint, the benefits of controlled and proper use of transgenics far exceed any risk. As of this date, no proven ill effects have been reported on human or animal health that were caused by the consumption of foods derived from the use of biotechnology, including genetic engineering (Ganeshan et al. 2008).

The possibilities of introducing target genes that influence herbicide and disease resistance into nontarget species by cross-pollination (gene flow) is a primary concern of environmentalists and agronomists who question the introduction of transgenic plants in agricultural production. Due to barley's self-fertilizing nature, it is considered a low-risk crop with regard to gene flow within the species genus (Ritala et al. 2002). Implementation of accepted and effective agricultural practices is a first step in reducing the risk of gene flow that could result in transfer of the target gene to nontarget plants. There are many plants (weeds) that are resistant to one or more chemical herbicides; there have also been crops bred by conventional methods to be resistant to herbicides. Do not these resistant plants also pose a threat or risk of transferring their resistance to other plants? Perhaps yes, as risk is always inherent with any new technology. As with the health safety issues, the low risk of environmental damage by transgenic barley can be managed effectively, but as with plants created by classical breeding, zero risk cannot be achieved. The issue boils down to making choices between risks and benefits.

The use of transgenics has the potential to increase productivity, quality, and uses of barley and other crops while making agriculture more environmentally friendly. Built-in genetic resistance to herbicides and diseases is a first step in making agriculture more environmentally friendly by reducing or eliminating the need for spraying with certain more noxious chemicals now used to control weeds and disease. According to Tourte (2005), "we can reasonably expect a development of the biotechnological approach in agronomic research with the triple objectives of a more productive, less costly, and more environmentally protective agriculture." Perhaps some day, traditional agriculture and "organic" agriculture may merge into one production system.

SUMMARY

In the past 100 years, great strides have been made in plant breeding technology. Traditional methods of plant breeding have been established and improved upon further as scientists gained greater depth in understanding genetic control in barley. Scientific advances in breeding techniques have not overshadowed the basic methods of improving the barley crop. The human element involved in the selection of parent barleys is recognized as being an all-important part of a breeding program. Also, a breeder must have a through knowledge of barley genetics, morphology, as well as knowledge of the proposed use of the end product. For many years, total crop yield was the single most important characteristic in breeding for improved cultivars. Yield is a quantitative characteristic controlled by multiple alleles that may or may not be advantageous to the end use of the grain. Thus, in later times, breeders have given more attention to end-product utilization, such as malt quality as well as yield, although the latter is still the single most important selection entity. Although acceptance of genetically engineered barley has not become a reality, due to negative consumer pressure, considerable research has been reported on the use of this technology, and further research is in progress. Introduction of genetic material into barley via genetic engineering for the purposes of improving crop agronomics and current numerous nutritional traits will eventually become a reality.

It is unfortunate that GMOs have received such a deluge of negative opposition from a very vocal minority. Much of the negativism has arisen from fear of the unknown and unwarranted beliefs that otherwise commonly accepted alleles will produce dire effects when expressed following introduction in a foreign genome. B. M. Chassey (2007) recently wrote an extremely informative and provocative essay on the history and future of GMOs in food and agriculture. (Readers who may have reserved feelings toward GMO foods as well as supporters of this technology are encouraged to read this article in *Cereal Food World*, July–August 2007, pp. 169–172.) A quote from Chassey's article is worthy of special note: "Simply put, the perceived hazards of transgenic technology are fiction." In the same issue of *Cereal Foods World*, Richard Stier, a consulting food scientist, echoed Chassey (Stier 2007), pointing out that it has been seven years since Frank-Roman Lauter wrote: "It is generally believed that the reluctance of consumers to accept bioengineered foods is without scientific basis. However, regulators and the agri-food industry must deal with the consumer concerns" (Lauter 2000). According to Stier, "the answer [to overcome barriers to use of bioengineered foods] is education of government regulators and the consuming public. The message needs to get out to the people and governments in a way that allows them to understand the technology, the safeguards that are built into it and the benefits."

REFERENCES

Aastrup, S. Erdal, K., and Munck, L. 1983. Low β-glucan barley mutants and their malting behavior. Pages 387–393 in: *Proc. 20th Congress of the European Brewery Convention*. Oxford University Press, Oxford, UK.

Anderson, M. K., and Reinbergs, E. 1985. Barley breeding. Pages 231–268 in: *Barley*. Agronomy Monograph 26. D. C. Rasmusson, ed. American Society of Agronomy, Crop Science Society of America, and Soil Science Society of America, Madison, WI.

Barley CAP (Barley Coordinated Agricultural Project). 2007. Published online at http:// www.BarleyCAP.org.

Borlaug, N. E. 1998. Feeding a world of 10 billion people: the miracle ahead. *Plant Tissue Cult. Biotechnol*. 3:119–127.

Bothmer, R. von. 1992. The wild species of *Hordeum*: relationships and potential use for improvement of cultivated barley. Pages 3–18 in: *Barley: Genetics, Biochemistry, Molecular Biology and Biotechnology*. P. R. Shewry, ed. C.A.B. International, Wallingford, UK.

Bothmer, R. von, Jacobsen, N., Baden, C., Jorgensen, R. B., and Linde-Laursen, I. 1995. An ecogeographical study of the genus *Hordeum*. Pages 1–19 in: *Systematic and Ecogeographic Studies on Crop Genepools*, vol. 7, 2nd ed. R. von Bothmer, ed. International Plant Genetic Research Institute, Rome.

Brettschneider, R., Lazzeri, P. A., Langridge, P., Hartke, S., Gill, R., and Lörz, H. 1990. Cereal transformation with *Agrobacterium tumefaciens* via the pollen tube pathway. Page 49 in: *Abstr. 7th International Congress of Plant Tissue and Cell Culture*. Abstract A2-22. Kluwer Academic, Dordrecht, The Netherlands.

Briggs, D. E. 1978. *Barley*. Chapman & Hall, New York.

Brinch-Pedersen, H., Galili, G., Knudsen, S., and Holm, P. B. 1996. Engineering of the aspartate family biosynthetic pathway in barley (*Hordeum vulgare* L.) by transformation with heterologeous genes encoding feed-back-insensitive aspartate kinase and dihydrodipicolinate synthase. *Plant Mol. Biol*. 32:611–620.

Chassey, B. M. 2007. The history and future of GMOs in food and agriculture. *Cereal Foods World* 52:168–172.

Cho, M.-J., Jiang, W., and Lemaux, P. 1998. Transformation of recalcitrant barley cultivars through improvement of regenerability and decreased albinism. *Plant Sci*. 138:229–244.

Cho, M.-J., Jiang, W., Lemaux, P. G., and Buchanan, B. B. 1999. Over expression of thioredoxin leads to enhanced activity of starch branching enzyme (pullulanase) in barley grain. *Proc. Natl. Acad. Sci. U.S.A*. 25:14641–14646.

Condon, R., Gustus, C., Rasmusson, D. C., and Smith, K. P. 2008. Effect of advanced cycle breeding on genetic diversity in barley breeding germplasm. *Crop Sci*. (in Press).

Cooper, L. D., Marquez-Cedillo, L., Singh, J., Sturbaum, A. K., Zhang, S., Edwards, V., Johnson, K., Kleinhofs, A., Rangel, S., Carollo, V., Bregitzer, P., Lemaux, P. G., and Hayes, P. M. 2004. Mapping *Ds* insertions in barley using a sequence based approach. *Mol. Genet. Genom*. 272:181–193.

CRIS USDA. 2007. Regional Barley Genome Mapping Project. Published online at http://www.cris.csrees.usda.gov.

Doll, H. 1973. Inheritance of the high-lysine character of a barley mutant. *Hereditas* 74:293–294.

Ellis, R. P. 2002. Wild barley as a source of genes for crop improvement. Pages 65–83 in: *Barley Science: Recent Advances from Molecular Biology to Agronomy of Yield and Quality*. G. A. Slafer, J. L. Molina-Cano, R. Savin, J. L. Araus, and I. Romagosa, eds. Haworth Press, Binghamton, NY.

Eslick, R. F. 1981. Mutation and characterization of unusual genes associated with the seed. Pages 864–867 in: *Barley Genetics IV: Proc. 4th International Barley Genetics Symposium*, Edinburgh, UK.

Eslick, R. F., and Hockett, E. A. 1979. Genetic engineering as a key to water use efficiency. *Agric. Meteorol*. 14:13–43.

FAO (Food and Agriculture Organization). 2007. Published online at http://www.faostat. fao.org/.

Fedak, G. 1985. Wide crosses in *Hordeum*. Pages 155–186 in: *Barley*. Agronomy Monograph 26. D. C. Rasmusson, ed. American Society of Agronomy, Crop Science Society of America, and Soil Science Society of America, Madison, WI.

Fedak, G. 1992. Intergeneric hybrids with *Hordeum*. Pages 45–70 in: *Barley: Genetics, Biochemistry, Molecular Biology and Biotechnology*. P. R. Shewry, ed. C.A.B. International, Wallingford, UK.

Ganeshan, S., Båga, M., Harvey, B. L., Rossnagel, B. G., Scoles, G. J., and Chibbar, R. N. 2003. Production of multiple shoots from thidiazuron-treated mature embryos and leaf-base/apical meristems of barley (*Hordeum vulgare* L.). *Plant Cell Tissue Organ Cult*. 73:57–64.

Ganeshan, S., Dahleen, L. S., Tranberg, J., Lemaux, P. G., and Chibbar, R. N. 2008. Barley. Pages in: *A Compendium of Transgenic Crop Plants*. C. Cole and T. C. Hall, eds. Wiley-Blackwell, Hoboken, NJ.

Guo, G., Maiwald, F., Lorenzen, P., and Steinbiss, H.-H. 1998. Factors influencing T-DNA transfer into wheat and barley cells by *Agrobacterium tumefaciens*. *Cereal Res. Communi*. 26:15–22.

Gürel, F., and Gözükermizi, N. 2002. Optimization of gene transfer into barley (*Hordeum vulgare* L.) mature embryos by tissue electroporation. *Plant Cell Rep*. 19:787–791.

Hagberg, A., and Hagberg, G. 1980. High frequency of spontaneous haploids in the progeny of an induced mutation in barley. *Hereditas* 93:341–343.

Harlan, H. V., and Martini, M. L. 1938. The effect of natural selection in a mixture of barley varieties. *J. Agric. Res*. 57:180–200.

Harlan, H. V., and Pope, M. N. 1922. The use and value of back-crosses in small-grain breeding. *J. Hered*. 7:319–322.

Hockett, E. A. 1981. Registration of hulless and hulless short-awned spring barley germplasm. *Crop Sci*. 21:146–147.

Hockett, E. A., and Eslick, R. F. 1968. Genetic male sterility in barley. I. Nonallelic genes. *Crop Sci*. 8:218–220.

Hockett, E. A., and Nilan, R. A. 1985. Genetics. In: *Barley* Agronomy Monograph 26. D. C. Rasmusson, ed. American Society of Agronomy, Crop Science Society of America, and Soil Science Society of America, Madison, WI, pp. 190–216.

Horsch, R. B., Fry, J. S., Hoffman, N. L., Eichholtz, D., Rogers, S. G., and Fraley, R. T. 1985. Simple and general methods for transferring genes in plants. *Science* 227:1229–1231.

Horvath, H., Huang, J., Wong, O. T., Kohl, E., Okita, T., Kannangara, C. G., and von Wettstein, D. 2000. The production of recombinant proteins in transgenic barley grains. *Proc. Nat. Acad. Sci. U.S.A*. 97:1914–1919.

Horvath, H., Huang, J., Wong, O. T., and von Westtstein, D. 2002. Experiences with genetic transformation of barley and characteristics of transgenic plants. Pages 143–175 in: *Barley Science: Recent Advances from Molecular Biology to Agronomy of Yield and Quality*. G. A. Slafer, J. L. Molina-Cano, R. Savin, J. L. Araus, and I. Romagosa, eds. Haworth Press, Binghamton, NY.

Islam, A. K. M. R., and Shepherd, K. W. 1990. Incorporation of barley chromosomes into wheat. Pages 128–151 in: *Biotechnology in Agriculture and Forestry: Wheat*, vol. 13. Springer-Verlag, Berlin.

Jacobsen, J. V., Matthews, P. M., Abbott, D. C., Wang, M. B., and Waterhouse, P. M. 2000. Transgenic barley. Pages 88–114 in: *Transgenic Cereals*. L. O'Brien and R. J. Henry, eds. American Association of Cereal Chemist, St. Paul, MN.

Jähne, A., Becker, D., Brettschneider, R., and Lörz, H. 1994. Regeneration of transgenic, microspore-derived, fertile barley. *Theor. Appl. Genet.* 89:525–533.

Jansson, C., and Ferstad, H-G. 2002. Mutants and transgenics: a comparison of barley resources in crop breeding. Pages 42–50 in: *Progress in Botany*, vol. 64. K. Esser, ed. Springer-Verlag, New York.

Jende-Strid, B. 1976. Mutations affecting flavonoid synthesis in barley. Page 36 in: *Barley Genetics III: Proc. 3rd International Barley Genetics Symposium*, Garching, Germany. H. Gaul, ed. Verlag Karl Thiemig, Munich, Germany.

Jensen, J. 1979. Location of a high-lysine gene and the DDT resistance gene of barley chromosome 7. *Euphytica* 24:47–56.

Jensen, L. G., Olsen, O., Kops, O., Wolf, N., Thomsen, K. K., and von Wettstein, D. 1996. Transgenic barley expressing a protein-engineered, thermostable $(1,3-1,4)$-β-glucanase during germination. *Proc. Nat. Acad. Sci. U.S.A.* 93:3487–3491.

Jensen, N. F. 1970. A diallel selective mating system for cereal breeding. *Crop Sci.* 10:629–635.

Jones, J. M. 1992. *Food Safety*. Eagan Press, St. Paul, MN.

Karp, A., and Lazzeri, P. W. 1992. Regeneration, stability and transformation of barley. Pages 549–571 in: *Barley: Genetics, Biochemistry, Molecular Biology and Biotechnology*. P. R. Shewry, ed. C.A.B. International, Wallingford, UK.

Kasha, K. J., and Kao, K. N. 1970. High frequency haploid production in barley (*Hordeum vulgare* L.). *Nature* 235:874–876.

Kasha, K. J., and Reinbergs, E. 1976. Utilization of haploidy in barley. Pages 307–315. in: *Barley Genetics III: Proc. 3rd International Barley Genetics Symposium*, Garching, Germany. H. Gaul, ed. Verlag Karl Thiemig, Munich, Germany.

Kasha, K. J., and Reinbergs, E. 1979. Achievements with haploids in barley research and breeding. Pages 215–230 in: *The Plant Genome: Proc. 2nd International Haploid Conference*. D. R. Davies and D. Hopwood, eds. John Inns Institute, Norwich, CT.

Kihara, M., Okada, Y., Kuroda, H., Saeki, K., Yoshigi, N., and Ito, K. 2000. Improvement of β-amylase thermostability in transgenic barley seeds and transgene stability in progeny. *Mol. Breed.* 6:511–517.

Klein, T. M., and Jones, T. J. 1999. Methods of genetic transformation: the gene gun. Pages 21–42 in: *Molecular Improvement of Cereal Crops*. I. K. Vasil, ed. Kluwer Academic, Dordrecht, The Netherlands.

Kleinhofs, A., and Han, F. 2002. Molecular mapping of the barley genome. Pages 31–63 in: *Barley Science: Recent Advances from Molecular Biology to Agronomy of Yield and Quality*. G. A. Slafer, J. L. Molina-Cano, R. Savin, J. L. Araus, and I. Romagosa, eds. Haworth Press, Binghamton, NY.

Kleinhofs, A., Chao, S., and Sharp, P. J. 1988. Mapping of nitrate reductase genes in barley and wheat. Pages 541–546 in: *Proc. 7th International Wheat Genetics Symposium*, vol. 1. T. E. Miller and R. M. D. Koebner, eds. Institute of Plant Science Research, Cambridge, UK.

Komari, T., and Kubo, T. 1999. Methods of genetic transformation: *Agrobacterium tumefaciens*. Pages 41–82 in: *Molecular Improvement of Cereal Crops*. I. K. Vasil, ed. Kluwer Academic, Dordrecht, The Netherlands.

Kruse, A. 1973. *Hordeum* × *Triticum* hybrids. *Hereditas* 73:157–161.

Lauter, F.-R. 2000. GMO labeling: the need for and status of GM-DNA quantification. *Cereal Foods World* 45:389–390.

Ledoux, O., and Huart, R. 1969. Fate of exogenous bacterial deoxyribonucleic acids in barley seedlings. *J. Mol. Biol.* 43:243–262.

Lemaux, P. G., Cho, M-J., Zang, S., and Bregitzer, P. 1999. Transgenic cereals: *Hordeum vulgare* L. (barley). Pages 255–316 in: *Molecular Improvement of Cereal Crops*. I. K. Vasil, ed. Kluwer Academic, Dordrecht., The Netherlands.

Luckett, D. J., and Darvey, N. L. 1992. Utilisation of microspore culture in wheat and barley improvement. *Aust. J. Bot.* 40:807–828.

Lupton, F. G. W., and Whitehouse, R. N. H. 1957. Studies on the breeding of self-pollinated cereals. *Euphytica* 6:169–184.

McCallum, C., Comai, L., Greene, E. A., and Henikoff, S. 2000. Targeted screening for induced mutations. *Nature Biotech*. 18:455–457.

Mejlhede, N., Kyjovska, Z., Backes, G., Burhenne, K., Rasmussen, S. K., and Jahoor, A. 2006. EcoTILLING for the identification of allelic variation in the powdery mildew resistance genes *mlo* and *mla* of barley. *Plant Breed.* 125:461–467.

Mendel, R. R., Claus, E., Hellmund, R., Schulze, J., Steinbiss, H. H., and Tewes, A. 1990. Gene transfer to barley. Pages 73–78 in: *Proc. 7th Int. Congress on Plant Cell and Tissue Culture*. Kluwer Academic, Dordrecht, The Netherlands.

Merritt, N. R. 1967. A new strain of barley with starch of high amylose content. *J. Inst. Brew. (London)* 73:583–585.

Morrell, M. K., Kosar-Hashemi, B., Cmiel, M., Samuel, M. S., Chandler, P., Rahman, S., Buleon, A., Batey, I. L., and Li, Z. 2003. Barley sex6 mutants lack starch synthase IIa activity and contain a starch with novel properties. *Plant J*. 34:173–185.

Munck, L. 1992. The case of high-lysine barley breeding. Pages 573–601 in: *Barley: Genetics, Biochemistry, Molecular Biology and Biotechnology*. P. R. Shewry, ed. C.A.B. International, Wallingford, UK.

Nevo, E. 1992. Origin, evolution, population, genetics, and resources for breeding of wild barley, *Hordeum spontaneum*, in the Fertile Crescent. Pages 19–43 in: *Barley: Genetics, Biochemistry, Molecular Biology and Biotechnology*. P. R. Shrewry, ed. C.A.B. International, Wallingford, UK.

Nuutila, A. M., Ritala, A., Skadsen, R. W., Mannonen, L., and Kauppinen, V. 1999. Expression of fungal thermotolerant endo-1,4-β-glucanase in transgenic barley seeds during germination. *Plant Mol. Biol*. 41:777–783.

Oerke, E.-C., Dehne, H.-W., Schöbeck, F., and Weber, A. 1994. *Crop Production and Crop Protection*. Elsevier, Amsterdam.

Park, S. J., Walsh, E. J., Reinbergs, E., Song, L. S. P., and Kasha, K. J. 1976. Field performance of doubled-haploid barley lines in comparison with lines developed by the pedigree and single-seed descent methods. *Can. J. Plant Sci.* 56:467–474.

Patel, M., Johnson, J. S., Brettell, R. I. S., Jacobsen, J., and Xue, G.-P. 2000. Transgenic barley expressing a fungal xylanase gene in the endosperm of developing grains. *Mol. Breed.* 6:113–123.

Pickering, R. A., and Devaux, P. 1992. Haploid production: approaches and use in plant breeding. Pages 519–547 in: *Barley: Genetics, Biochemistry, Molecular Biology and Biotechnology.* P. R. Shewry, ed. C.A.B. International, Wallingford, UK.

Pickering, R. A., and Johnston, P. A. 2005. Recent progress in barley improvement using species of *Hordeum. Cytogenet. and Genome Res.* 109:344–349.

Pickering, R., Hill, A. M., Michel, M., and Timmerman-Vaughan, C. M. 1995. The transfer of a powdery mildew-resistant gene from *Hordeum bulbosum* to barley (*H. vulgare* L.) chromosome 2(2I). *Theor. Appl. Genet.* 91:1288–1292.

Ramage, R. T. 1975. Techniques for producing hybrid barley. *Barley Newslett.* 18:62–65.

Ramesh, S. A., Choimes, S., and Schachtman, D. P. 2004. Over-expression of an *Arabidopsis* zinc transporter in *Hordeum vulgare* increases short-term zinc uptake after zinc deprivation and seed zinc content. *Plant Mol. Biol.* 54:373–385.

Rao, H. S., Basha, O. P., Singh, N. K., Sato, K., and Dhaliwal, H. S. 2007. Frequency distributions and composite interval mapping for QTL analysis in 'Steptoe' × 'Morex' barley mapping population. *Barley Genet. Newsl.* 37:520. Published online at http://www.wheat.pw.usda.gov/ggpages/bgn.

Rasmusson, D. C., and Phillips, R. L. 1997. Plant breeding progress and genetic diversity from de novo variation and elevated epistasis. *Crop Sci.* 37:303–310.

Reinbergs, E., Park, S. J., and Kasha, K. J. 1976. Haploid technique in comparison with conventional methods in barley breeding. Pages 346–346 in: *Barley Genetics III: Proc. 3rd International Barley Genetics Symposium, Garching, Germany.* H. Gaul, ed. Verlag Karl Thiemig, Munich, Germany.

Ritala, A., Apegren, K., Kurtén, U., Salmenkaillo-Marttila, Mannonen, L., Hannus, R., Kauppinen, V., Teeri, T. H., and Enari, T. 1994. Fertile transgenic barley by particle bombardment of immature embryos. *Plant Mol. Biol.* 24:317–325.

Ritala, A., Nuutila, A. M., Aikasalo, R., Kauppinen, V., and Tammisola, J. 2002. Measuring gene flow in the cultivation of genetic barley. *Crop Sci.* 42:278–285.

Shepherd, K. W., and Islam, A. K. M. R. 1992. Progress in the production of wheat–barley addition and recombination lines and their use in mapping the barley genome. Pages 99–114 in: *Barley: Genetics, Biochemistry, Molecular Biology and Biotechnology.* P. R. Shewry, ed. C.A.B. International, Wallingford, UK.

Singh, J., Zhang, S, Chen, C., Cooper, L., Bregitzer, P., Sturbaum, A. K., Hayes, P. M., and Lemaux, P. G. 2006. High-frequency reactivation of maize *Ds* favors saturation mutagenesis in barley. *Plant Mol. Biol.* 62:937–950.

Slade, A. J., Fuerstenberg, S. I., Loeffler, D., Steine, M. N., and Facciotti, D. 2005. A reverse genetic, nontransgenic approach to wheat crop improvement by TILLING. *Nat. Biotechnol.* 23:75–81.

Soyfer, V. N. 1980. Hereditary variability of plants under the action of exogenous DNA. *Theor. Appl. Genet.* 107:958–964.

Starling, T. M. 1980. Barley. Pages 189–201 in: *Hybridization of Crop Plants*. W. R. Fehr and M. M. Madley, eds. American Society of Agronomy, Madison, WI.

Stier, R. 2007. GMOs: manna from heaven or frankenfood? *Cereal Foods World* 52:207.

Swanston, J. S., and Ellis, R. P. 2002. Genetics and breeding of malt quality attributes. Pages 85–114 in: *Barley Science: Recent Advances from Molecular Biology to Agronomy of Yield and Quality*. G. A. Slafer, J. L. Molina-Cano, R. Savin, J. L. Araus, and I. Romagosa, eds. Haworth Press, Binghamton, NY.

Thomas, T. B. 2002. Molecular marker-assisted versus conventional selection in barley breeding. Pages 177–204 in: *Barley Science: Recent Advances from Molecular Biology to Agronomy of Yield and Quality*. G. A. Slafer, J. L. Molina-Cano, R. Savin, J. L. Araus, and I. Romagosa, eds. Haworth Press, Binghamton, NY.

Till, B. J., Reynolds, S. H., Greene, E. A., Codomo, C. A. Enns, L. C., Johnson, J. E., Burtner, C., Odden, A. R., Young, K., Taylor, N. E., Henikoff, J. F. G., Comai, L., and Henikoff, S. 2003. Large-scale discovery induced point mutations with high-throughput TILLING. *Genome Res.* 13:524–530.

Tingay, S., McElroy, D., Kalla, R., Fieg, S., Wang, M., Thornton, S., and Brettell, R. 1997. *Agrobacterium tumefaciens*–mediated barley transformation. *Plant J.* 11:1369–1376.

Töpfer, R., Gronenborn, B., Schell, J., and Steinbiss, H. H. 1989. Uptake and transient expression of chimaeric genes in seed-derived embryos. *Plant Cell Rep.* 1:133–139.

Tourte, Y. 2005. *Genetic Engineering and Biotechnology: Concepts, Methods, and Agronomic Applications*. Science Publishers, Enfield, NH.

Tull, D., Phillipson, B. A., Kramhoft, B., Knudsen, S. Olsen, O., and Svensson, B. 2003. Enhance amylolytic activity in germinating barley through synthesis of a bacterial alpha-amylase. *J. Cereal Sci.* 37:71–80.

Turcotte, P., St. Pierre, C. A., and Ho, K. M. 1980. Comparison entre des lignées pedigrées et des lignées haploïds doubles chez l'orge (*Hordeum vulgare* L.). *Can. J. Plant Sci.* 60:79–85.

Ullrich, S. E. 2002. Genetics and breeding of barley feed quality attributes. Pages 115–142 in: *Barley Science: Recent Advances from Molecular Biology to Agronomy of Yield and Quality*. G. A. Slafer, J. L. Molina-Cano, R. Savin, J. L. Araus, and I. Romagosa, eds. Haworth Press, Binghamton, NY.

Vasil, I. K. 1999. Molecular improvement of cereal crops: an introduction. Pages 1–8 in: *Molecular Improvement of Cereal Crops*. I. K. Vasil, ed. Kluwer Academic, Dordrecht, The Netherlands.

Wan, Y., and Lemaux, P. G. 1994. Generation of large numbers of independently transformed fertile barley plants. *Plant Physiol.* 104:37–48.

Wiebe, G. A. 1978. Breeding. Pages 117–127 in: *Barley: Origin, Botany, Culture, Winter Hardiness, Genetics, Utilization, and Pests*. Agriculture Handbook 338. U.S. Department of Agriculture, Washington, D.C.

Wu, H., McCormac, A. C., Elliot, M. C., and Chen, D.-F. 1998. *Agrobacterium*-mediated stable transformation of cell suspension cultures of barley (*Hordeum vulgare*). *Plant Cell Tissue Organ Cult.* 54:161–171.

Xue, G. P., Patel, M. Johnson, J. S., Smyth, D. J., and Vickers, C. E. 2003. Selectable marker-free transgenic barley producing a high level of cellulase (1,4-β-glucanase) in developing grains. *Plant Cell Rep.* 21:1088–1094.

Zhang, S., Cho, M.-J., Koprek, T., Yun, R., Bregitzer, P., and Lemaux, P. G. 1999. Genetic transformation of commercial cultivars of oat (*Avena sativa* L.) and barley (*Hordeum vulgare* L.) using in vitro shoot meristematic cultures derived from germinated seedlings. *Plant Cell Rep.* 18:959–966.

Zhang, Y., Darlington, H., Jones, H. D., Halford, N. G., Napier, Davey, M. R., Lazzeri, P. A., and Shewry, P. R. 2003. Expression of the gamma-zein protein of maize in seeds of transgenic barley: effects on grain composition and properties. *Theor. Appl. Genet.* 106:1139–1146.

4 Barley: Genetics and Nutrient Composition

INTRODUCTION

Information presented in Chapter 2 describes the complexity of the barley plant and kernel, the growth of which is controlled by the barley genome from germination to maturity. The inheritance of certain characteristics is critical to the problem of breeding barley for food or any other desired end uses. Perhaps the importance of the genetic code is emphasized when it is recognized that all cultivated barley used for feed, malt, or food is dependent on the tough rachis that is controlled by the *brittle* genes, *Btr1* and *Btr2*, found on chromosome 3(3H). Cultivated barleys are homozygous recessive for either one or the other gene or for both genes that produce a tough rachis (Harlan 1978). Without a tough rachis, harvesting mature barleys seeds would be difficult because this characteristic prevents scattering of seeds by any number of natural events, such as wind or rain.

There are several characteristics under genetic control that can be introduced selectively into barley to influence or dictate nutrient composition. Additionally, genetic selection can be employed to incorporate value-added attributes such as ease of harvesting or processing, product appearance, taste, and health-promoting factors of barley foods. Although it is not possible to control completely the level and occurrence of all nutrients in barley, certain genes can be inserted or expressed in the genome, producing minor to major influences on composition.

Nutrients are unevenly dispersed in the various parts of the kernel, some being more concentrated in one area while others are concentrated in other parts. The location and concentration of the nutrients in the kernel have major implications in nutrient availability, processing methods, and in food preparation. Nutritional parameters of barley are generally reported as averages, when in reality barleys differ as much or more in nutritional characteristics as in morphological characteristics. The amounts and types of the various nutrients are controlled by

Barley for Food and Health: Science, Technology, and Products,
By Rosemary K. Newman and C. Walter Newman
Copyright © 2008 John Wiley & Sons, Inc.

genetics and by environmental growing conditions, including farming practices such as irrigation and genetic–environment interaction. By using superior germplasm and controlling certain environmental factors such as moisture available through irrigation, a reasonably uniform barley crop can be produced from one growing season to the next. In this chapter we provide a brief introduction to the genetics of barley, the application of genetics in breeding barley for food, and the chemical components or nutrients of the barley kernel relative to their location in the kernel.

GENETICS AND NUTRIENT COMPOSITION

To date, a great amount of scientific information on the barley genome has been generated. This is partially due to the international importance of barley as a major crop but also to the dedication of scientists who have a genuine interest in studying this fascinating grain. The genetics of barley has probably been explored to a greater extent than that of any other cereal grain. According to Hockett and Nilan (1985), cultivated and wild barleys are major experimental organisms among flowering plants for genetic studies. The wide use of barley in genetic studies was attributed to (1) its diploid nature, (2) its low chromosome number ($2n = 14$), (3) the relatively large chromosomes (6 to 8 μm), (4) the high degree of self-fertility, and (5) the ease of hybridization. The reader is referred to the following for discussion of barley genetics in greater depth than is presented in this book: Smith 1951; Nilan 1964, 1974; Hockett and Nilan 1985; Wettstein-Knowels 1992; Nilan and Ullrich 1993; Barley Genetics Newsletter 1996; Kleinhofs and Han 2002; and Ullrich 2002.

The barley genome contains a vast number of genes that determine food quality traits, many of which have not yet been characterized. Some traits are unique to barley, but most are shared with other cereal grain species. There are several anatomical traits that are controlled by one or two genes (simply inherited), but most traits involving complex physiological processes are inherited quantitatively, being controlled by several to many genes. Genetic knowledge of many simply inherited traits dates back many years, whereas reliable data about the inheritance of the quantitative traits have been limited until relatively recently with the development of comprehensive molecular marker–based genetic maps and the technology of *quantitative trait locus* (QTL) *analysis* (Ullrich 2002). The importance of this technology was summed up by Nilan (1990): "The comprehensive molecular marker linkage maps have proved a powerful tool in barley for identifying QTL loci for agronomic and quality traits, determining QTL effects and action, and facilitating genetic engineering."

In discussing genetically controlled traits in a biological system such as barley, it is customary to identify the genes(s) and the location on the chromosome involved when this information is known. Barley has seven pairs of chromosomes that have been defined based on their sizes and characteristics (Nilan 1964; Ramage 1985). Chromosomes 1 through 5 differ in their sizes measured at mitotic metaphase, with chromosome 1 being the longest and chromosome 5 the shortest. Chromosomes 6 and 7 have satellites, with chromosome 6 having the larger satellite of the two (Kleinhofs and Han 2002). The barley genome is well characterized, with over 1000 genes and 500 translocation stocks known (Sogaard and Wettstein-Knowles 1987; Wettstein-Knowles 1992). Conventional genetic mapping over the past 50 years has placed over 200 loci to the barley chromosomes (Franckowiak 1997). Recently, comparative mapping studies have revealed that barley chromosomes numbers 1, 2, 3, 4, 5, 6, and 7 are homologous to wheat chromosomes 7H, 2H, 3H, 4H, 1H, 6H, and 5H, respectively. Current recommendations are that barley chromosomes be designated according to their homologous relationships with chromosomes of other Triticeae species (Linde-Laursen 1997). It has been suggested that the barley chromosome designations be used with the Triticeae designations in parentheses to avoid confusion with what has been published previously (Kleinhofs and Han 2002).

Nutritional components of barley are generally reported as averages, when in reality, barley may differ greatly in chemical composition due to genotype, cultural practices, and environmental growing conditions. Extreme variability can be demonstrated when experimental and unconventional barley types are included in a comparison (Newman and McGuire 1985). However, the barley nutrient composition values that are generally encountered in commerce are more comparable to average values. A composite of the major components of a barley kernel are shown in Table 4.1. Starch, fiber, and protein make up the largest portion of the kernel, and variation in one of these components will influence the amounts of the other two. Starch level varies inversely with protein ($r = -0.81$) and fiber ($r = -0.64$) (Åman and Newman 1986). Nutritional groups that will be considered in this discussion are carbohydrates (starch, sugars, and fiber), protein, lipids, vitamins, minerals, and phytochemicals.

TABLE 4.1   Typical Composition (g/kg) of Hulled and Hulless Isotype Barleys, Dry Matter Basis

Item	Hulled		Hulless	
	Mean[a]	Range	Mean[a]	Range
Protein[b]	13.7	12.5–15.4	14.1	12.1–16.6
Starch	58.2	57.1–59.5	63.4	60.5–65.2
Sugars[c]	3.0	2.8–3.3	2.9	2.0–4.2
Lipids	2.2	1.9–2.4	3.1	2.7–3.9
Fiber	20.2	18.8–22.6	13.8	12.6–15.6
Ash	2.7	2.3–3.0	2.8	2.3–3.5

Source: Adapted from Åman and Newman (1986).
[a] $n = 3$.
[b] $N \times 6.25$.
[c] Glucose, fructose, sucrose, and fructans.

TABLE 4.2   Carbohydrates (g/kg) and Extract Viscosity of Nonwaxy and Waxy Barley Genotypes, Dry Matter Basis

Item[a]	Nonwaxy		Waxy	
	Hulled	Hulless	Hulled	Hulless
Starch	55.9	61.3	51.5	58.5
Sugars[b]	2.3	2.9	5.0	5.5
Fiber				
Total dietary fiber	17.0	13.2	19.6	13.8
Soluble dietary fiber	4.4	4.9	5.9	6.3
Arabinoxylans	6.2	4.4	6.7	4.6
Cellulose	3.8	2.1	4.4	1.9
Lignin	2.0	0.9	1.8	0.9
Total β-glucans	4.4	4.7	5.3	6.3
Soluble β-glucans	2.6	2.6	3.2	3.4
Extract viscosity (cP)	2.8	3.1	3.3	4.9

Source: Adapted from Xue (1992).
[a] $n = 6$ for hulled types and $n = 12$ for hulless types.
[b] Glucose, fructose, sucrose, maltose, and fructans.

Carbohydrates

Carbohydrates are described in a number of ways, some of which are quite general and have little meaning or application to foods and nutrition. Broadly speaking, carbohydrates are composed of carbon, hydrogen, and oxygen, but in some instances in nature they are complexed with nitrogenous or lipid compounds.

Large-molecular-weight compounds such as starch and cellulose are referred to as *complex carbohydrates*, as opposed to simple sugars such as glucose, fructose, and sucrose. Complex carbohydrates, which make up the bulk of grain carbohydrates, are generally termed *polysaccharides*, which in barley and other cereals are divided into starch (soluble polysaccharides) and nonstarch or insoluble polysaccharides. The major nonstarch polysaccharides of barley include mixed linked (1,3)(1,4)-β-D-glucans (β-glucans) and arabinoxylans. These compounds, along with small amounts of (1,4)-β-D-glucan (cellulose), glucomannan, and (1,3)-β-glucan inclusively, are referred to as *total dietary fiber* (TDF). Carbohydrates as a whole are the largest component of the barley kernel, making up approximately 80% of the total dry matter. Although not a carbohydrate, lignin is generally included as a part of the carbohydrate complex because it is intimately associated with the fiber components, especially cellulose (Table 4.2).

Starch

Most soluble carbohydrates are deposited and stored within cells in the starchy endosperm tissue during the maturation of the kernel. The largest component of the carbohydrate group is starch, considered a soluble polysaccharide and the major source of energy in barley for food and growth of the new plant after germination. Starch is also the most variable constituent of all nutrients in barley, varying from a low of 45% or less to as much as 65% of the kernel, due primarily to environmental growing conditions. However, certain barley cultivars have 50 to 75% less starch than the average barley, due to incomplete filling of the endosperm. This is a genetically controlled factor that effectively removes a starch-synthesizing enzyme system producing a shrunken endosperm (Eslick 1981; Morell et al. 2003). In one instance the barley is waxy (high amylopectin) (Eslick 1981), and in the other, high in amylose (Morell et al. 2003). The waxy type has been reported to contain high levels of sugars (Hofer 1985; Åman and Newman 1986).

Barley starch is composed of two structural types, amylopectin and amylose. Amylopectin is a branched-chain polysaccharide in which α-(1,4)-D-glucose units are branched through α-(1,6)-D-glucose linkages, and amylose is primarily a straight chain of α-(1,4)-D-glucose units (Briggs 1978). Although amylose is generally considered the linear component of starch, there is limited branching in the molecule (MacGregor and Fincher 1993). The side chains have not been characterized completely in barley amylose, but may range in size from four to more than 100 glucose residues (Takeda et al. 1984). Barley amylopectin is a large molecule (Banks et al. 1972; DeHaas and Goering 1972), but despite its large size has a relatively low viscosity in dilute solution compared to that of amylose. Barley amylose molecules are smaller than amylopectin molecules, existing in a random coil structure. Amylose is unstable in aqueous solutions, tending to precipitate or retrograde in dilute solutions and form gels in concentrated solutions (MacGregor and Fincher 1993).

In most barley, amylopectin is the major starch type, making up 72 to 78% of the total, with amylose making up the remaining 22 to 28%. Although barley most often contains amylopectin and amylose in the 3:1 ratio, certain cultivars contain starch that is 95 to 100% amylopectin, and there are at least three cultivars that contain 40 to 70% amylose as a percentage of the total starch. The term *waxy barley* is applied to cultivars that contain high levels of amylopectin, and *nonwaxy barley* is applied to barleys containing starch with the normal 3:1 ratio of amylopectin to amylose. The *wax* gene, located on chromosome 1(7H), conditions a high level ($\approx$ 95%) of amylopectin starch and is recessive to the gene *Wax*, which is responsible for the 3:1 amylopectin/amylose ratio (Nilan 1964). Waxy barleys containing 100% amylopectin starch have been developed in conventional cross-breeding programs in Canada (Bhatty and Rossnagel 1997) and Sweden (Ajithkumar et al. 2005), and in Japan through chemical mutation using sodium azide (Ishikawa et al. 1995). Azhul, a waxy hulless barley, was the source of the waxy gene in the Canadian and Swedish reports. It is fairly well accepted that waxy barleys contain 5 to 8% less total starch than comparable nonwaxy types (Ullrich et al. 1986; Xue et al. 1997). However, the reduced level of starch is usually accompanied by small increases in the simple sugars fructose, glucose, and sucrose and the fiber component β-glucan (Xue et al. 1997).

Barleys that contain more than the normal amounts of amylose (40 to 70% of the total) in the starch are classified as high-amylose cultivars. The first reported high-amylose barley was a mutant of Glacier, a six-rowed feed type, and was designated Glacier Ac38. This barley was found to contain 44 to 47% amylose in the starch (Merritt 1967), and Walker and Merritt (1969) subsequently demonstrated that the amylose level in Glacier Ac38 was under genetic control. The controlling gene, *amol*, is located on chromosome 5(1H) (Barley Genetics Newsletter 1996). Morell et al. (2003) conducted a study on shrunken endosperm (sex6) barleys derived from Himalaya, a hulless nonwaxy variety. They identified two lines, M292 and M342, having high-amylose phenotypes resulting from a decrease in amylopectin synthesis. M292 and M342 were found to have amylose contents of 71.0 and 62.5% of the total starch, respectively. The starch properties of these two lines are unique in that there is an increase in short chains of amylopectin and the starch has a low gelatinization temperature. Additionally, starch granules of M292 and M342 have reduced levels of crystallinity, and the crystal shifts from A-type to a mixture of V- and B-types. *V-type starch* is typical of amylose–lipid complexes, indicating that a portion of the amylose in these barleys is complexed with lipid, which is highly unusual in cereal starches. Genetic analysis revealed that the gene causing the mutation produced impaired amylopectin synthesis in these barleys and is located at the *sex6* locus on chromosome 7H. Additionally, it was reported that sex6 starches have shortened amylopectin chain length compared to the high-amylose starch produced by the *amol* gene in Glacier Ac38.

Barley starch consists of a mixture of large-diameter (15 to 25 μm) granules and irregularly shaped granules that are smaller in diameter (<10 μm). The large and small granules are referred to as *A-* and *B-type granules*, respectively. Small granules, 2 to 4 μm in diameter, make up 80 to 90% of the total number of starch granules but only 10 to 15% of the total starch weight. Large granules measuring 10 to 30 μm are smaller in number but represent 85 to 95% of the total weight of the starch (Goering et al. 1973). Starch granule sizes in high-amylose and waxy barleys differ considerably from those found in normal barley starch. B-type granules are larger and A-type granules are smaller in high-amylose barley starch than in normal barley starch (Banks et al. 1973; Morrison et al. 1986). The granules in high-amylose barley starch have a more uniform size distribution, as the small granules fraction makes up a relatively high proportion of the total. In waxy barley starch, small granules are similar in size to small granules in normal barley starch, and as in the latter, the large granules occur in smaller numbers but constitute a high-weight fraction of the total (Morrison and Scott 1986). Goering et al. (1973) stated that the large granules in high-amylose and waxy barley starches appeared to have fewer granules in the range 20 to 30 μm than are found in normal barley starch. Starch is stored exclusively in the endosperm of the mature kernel, but is not distributed uniformly. Centrally located endosperm cells contain greater levels of starch than cells in the subaleurone, which preferentially contain more protein than starch. The precise way in which amylopectin and amylose are arranged in normal barley starch is unknown (MacGregor and

Fincher 1993). The type and amount of starch in barley have varying effects on the nutritional quality, processing characteristics, and end-product utilization.

Sugars

Low levels of simple sugars and oligosaccharides (small polymers) are found in the barley kernel in addition to the starch components (Henry 1988). Glucose and fructose (monosaccharides) are the simplest carbohydrates, occurring at very low levels (<0.2%) and are found mainly in the mature endosperm. Small amounts of the disaccharide maltose (0.1 to 0.2%) are sometimes found in barley endosperm tissue near the embryo (MacLeod 1952). The small amounts of maltose reported are thought to be the results of amylolytic enzyme activity, as maltose is not in the starch synthesis pathway. Higher levels of maltose (0.4%) have been reported in waxy barleys than in nonwaxy types (Xue et al. 1997), which may have been due to the presence of active α-amylase complexed with the starch in the waxy types (Goering and Eslick 1976). The disaccharide sucrose, which is a component in the starch synthesis pathway, represents 50% of the simple sugars found in barley kernels. Levels of sucrose ranging from 0.74 to 0.84% have been reported, with 80% of this located in the embryo (MacGregor and Fincher 1993). As stated in the discussion above on waxy barleys, mutant waxy barleys have been reported to contain up to 7.0% sucrose (Hofer 1985; Åman and Newman 1986). The trisaccharide raffinose has been reported at levels ranging from 0.3 to 0.8%, and as with sucrose, 80.0% of this sugar is found in the embryo (MacLeod 1952, 1957). Fructo-oligosaccharides (fructans) are polymers of fructose residues linked to a terminal glucose residue. Fructans containing 10 fructosyl residues have been identified in barley kernels (MacLeod 1953).

Waxy barley can easily be identified by testing with a solution of potassium iodide. The procedure is to cut the kernel, exposing the endosperm and staining it with the iodide solution. Normal and high-amylose endosperms will stain dark blue, and waxy endosperm will stain dark red or brownish (MacGregor and Fincher 1993). High-amylose barley can be identified by observing isolated starch granule size with a low-power microscope. Compared to normal or waxy starches, high-amylose starch will have a preponderance of small granules (D. R. Clark, personal communication).

Nonstarch Polysaccharides

The nonstarch polysaccharides (NSPs) of the barley kernel are structural components of the cell walls in hull, aleurone, and endosperm tissue. Proteins and glycoproteins form a secondary fibrous network within the matrix phase of the NSPs. Collectively, the NSPs, some of which have major impacts on end-product use, both positive and negative, comprise the total dietary fiber (TDF) in barley. In contrast to starch and sugars, NSPs are not digested by the human digestive system, thus provide little or no energy, but are valuable diet constituents for other reasons. Although not broken down by mammalian enzymes, microbial digestion of NSPs occurs in the large bowel, producing a number of breakdown products, including the short-chain fatty acids: acetic, propionic, and butyric acids. The amounts of the NSP components vary in concentration in the different tissues of the kernel.

The major NSPs in barley are β-glucans, arabinoxylans, and cellulose. Cellulose is a long-chain polymer of (1,4)-β linked glucose residues found primarily in the hull, with small amounts located in the aleurone and starchy endosperm. Lignin is closely associated with the cellulose in the hull. Arabinoxylan, a polymer of the pentose sugars arabinose and xylose, and β-glucans are integral components of the cell wall structure in both aleurone and starchy endosperm tissue but vary inversely in their ratios in the two tissues. Arabinoxylan and β-glucan make up 67 and 26% of the aleurone cell walls and 20 and 70% of the starchy endosperm cell walls, respectively (Duffus and Cochrane 1993). Small amounts of glucomannan and (1,3)-β-glucans are found in the aleurone and endosperm cell walls. Oats and barley are the only cereal grains in which significant amounts of β-glucans are found. As opposed to oats, in which the β-glucans are located principally in the outer portion of the kernel, barley β-glucan is distributed throughout the aleurone and endosperm as a structural cell wall component in most barley (Miller and Fulcher 1994; Zheng et al. 2000; Yeung and Vasanthan 2001).

Cellulose is a large-molecular-weight glucan that contains an immense number of β-glucosyl residues joined through β-(1, 4) linkages to form long, linear, ribbonlike chains. In addition to being insoluble in aqueous solutions, cellulose molecules are relatively inflexible, due to hydrogen bonding of adjacent glucosyl residues (Gardner and Blackwell, 1974). Alignment and aggregation of individual cellulose chains produce crystalline microfibrils. The primary site of cellulose in the barley kernel is in the hull, making up about 40% of the hull's dry weight. Over 96% of the total grain cellulose is present in the hull (MacLeod 1959), with very little present in the aleurone and endosperm and none in the embryo (Fincher and Stone 1986). Cellulose levels were reported to range from 4.1 to 4.8% in hulled barleys and 2.0 to 2.9% in hulless barleys (Xue et al. 1997). Lignin, silica, and arabinoxylans are intimately involved with cellulose microfibrils in the hull (Aspinall and Ross 1963), affording physical support for the embryo and the endosperm. Additionally, the aggregate structure of cellulose, lignin, and silica affords protection against insect and microbial penetration and possibly reduces desiccation under limited moisture conditions (Fincher and Stone 1986). The hull also provides protection for the coleoptile during germination, being especially important for this in the malting process. Cellulose microfibrils, constituting about 2% of the cell walls in barley aleurone and starch endosperm, are apparently embedded in a β-glucan and arabinoxylan matrix. Extensive intermolecular hydrogen bonding stabilizes the microfibrils, giving strength to the cell walls.

Barley cell wall arabinoxylans are members of a polysaccharide family in which member compounds differ in molecular size, composition, and structure and solubility (Viëtor et al. 1993). These polysaccharides, composed principally of arabinose and xylose, are found primarily with β-glucans in barley aleurone and endosperm cell walls, but also occur in the hull (Aspinall and Ross 1963).

In the endosperm they comprise 20 to 25% of the cell wall material and 86% in the aleurone. Barley arabinoxylans vary considerably with respect to the ratio of xylose to arabinose. In hulls, ratios of about 1 : 9 have been observed (Aspinall and Ferrier 1957), whereas in starchy endosperm or aleurone tissues, ratios of 1 : 3 were reported in a number of studies reviewed by MacGregor and Fincher (1993). From the literature it appears that the ratios reported are very dependent on the solvent used to solubilize the molecules. Arabinoxylans found in barley generally have a (1-4)-β-xylopyranosyl backbone that supports one or two α-L-arabinofuranosyl residues. Most arabinose is found as monomeric substituents but with a small proportion of oligomeric sidechains, consisting of two or more arabinosyl residues or an arabinosyl residue with a terminal xylosyl residue (Voragen et al. 1992; Viëtor et al. 1993). Galactosyl, glucosyl, and xylosyl can be present as terminal residues, but are quantitatively minor; however, glucuronopryanosyl terminal residues may constitute up to 4% by weight of the arabinoxylan in the hull (Aspinall and Ross 1963).

Estimates of arabinoxylan contents of barley grain range from 4 to 7% by weight, being concentrated in the outer layers of the caryopsis and the hull (Hashimoto et al. 1987). It was reported by Henry (1987) that less than 25% of total barley arabinoxylan is found in the endosperm, constituting about 1.5% by weight. Henry (1986) concluded that total arabinoxylan content and xylose–arabinose ratios varied somewhat with genotype but that content varied more, due to environmental factors. For example, thin or small kernels with low starch content contain higher amounts of arabinoxylans than large or plump kernels containing high levels of starch. This is presumed to be because most arabinoxylan is located in the outer layers of the grain, and small thin grains contain relatively less endosperm than plump large grains. According to MacGregor and Fincher (1993), there have been no comprehensive studies on the genetic basis for arabinoxylan levels in barley, nor is there information on the number of genes involved or their location in the genome.

Of all the components of TDF in barley, β-glucans are probably the most important in terms of human diet and health benefits. β-Glucans consist of high-molecular-weight linear chains of β-glucosyl residues polymerized through both β-(1,3) and β-(1,4) linkages. Barley β-glucans are members of a family of polysaccharides that is heterogeneous with respect to molecular size, solubility, and molecular structure (MacGregor and Fincher 1993). Although it is not possible to show a single structure for β-glucan, it has been demonstrated that blocks of two or three contiguous (1,4)-linked β-glucosyl residues are separated by single β-(1,3) linkages along the polysaccharide chain (Parrish et al. 1960; Woodward et al. 1983a). The β-glucosyl units are referred to as cellotriosyls (trisaccharides) and cellotetraosyls (tetrasaccharides). A schematic representation of barley β-glucan is shown in Figure 4.1.

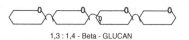

1,3 : 1,4 - Beta - GLUCAN

FIGURE 4.1 β-Glucan structure.

Unlike cellulose, β-glucans are partially soluble in aqueous solutions, a characteristic that is attributed to the molecular structural differences of the two polysaccharides. The presence of the interspersed β-(1,3) bonding creates a molecule with quite different spatial arrangements compared to total β-(1,4) bonding, which no doubt relates to the difference in solubility between the two polysaccharides. The irregularly spaced β-(1,3)-linkages interrupt the relatively rigid ribbonlike β-(1,4)-glucan conformation as in cellulose and confer flexibility and irregular shape on the barley β-(1,3)(1,4)-glucan, consistent with its water solubility (Woodward et al. 1983b). The solubility of barley β-glucans was first recognized and utilized by the brewing industry in identifying desirable cultivars for malting. Barley β-glucans are regarded as undesirable components in the malting and brewing processes for three reasons: (1) high levels of β-glucans in malt are associated with lower malt extract values; (2) β-glucans produce viscous extracts, causing difficulty in filtration; and (3) undegraded β-glucans can create haze or precipitates in beverages (Woodward and Fincher 1983; Bamforth 1985). The viscosity of β-glucan extracts can be used under laboratory conditions to indicate the level of β-glucans in barley. Thus, this characteristic was utilized in early malt barley breeding programs to estimate the level of β-glucans in barleys (Greenberg and Whitmore 1974; Bendelow 1975; Morgan 1977). Under laboratory conditions, the extract viscosity is altered not only by the β-glucan level but also by the experimental conditions (i.e., solvent pH, temperature, and concentration) and by the instruments used in recording the viscosity measurements. Viscosities of pure aqueous extracts are reduced rapidly due to native β-glucanase activity and are therefore not generally used. Most often, acid or alkaline extracts that deactivate the enzyme system are used to avoid this confounding factor. Regardless of method utilized, genotypic differences in extract viscosity correspond to differences in β-glucan content, with the exception of one or two cultivars (Fastnaught 2001). The viscosity of soluble barley β-glucans has negative implications for the malting and brewing industries and in some instances food production, but is considered an advantage in dietary and health applications (Bhatty 1992).

The β-glucan level in barley is under genetic control, although the concentration is often modified by the environment, especially hot, dry conditions during kernel maturation. Such conditions almost always produce increased levels of β-glucans (Bendelow 1975; Anderson et al. 1978). It was suggested that the increase in β-glucans may be caused by either a decrease in grain filling due to impaired starch and/or protein synthesis or because β-glucan synthesis is enhanced in such an environment. On the other hand, under moist conditions, such as rain during ripening, kernel β-glucan levels are decreased (Bendelow 1975; Aastrup 1979). Regardless of the fact that barley β-glucan levels are significantly influenced by environmental factors, there is general agreement among researchers that genetic background is the most important factor of final β-glucan content of the barley kernel (MacGregor and Fincher 1993). However, the genetics of barley kernel β-glucan has proven to be complex and is not completely understood (Ullrich 2002). Some specific genetic controls of β-glucan concentrations along with linkages to genes controlling other components have been identified. Aastrup (1983) and Molina-Cano et al. (1989) identified low-β-glucan mutant barleys that were attributed to a simple inheritance factor. Using crossbreeding systems, Greenberg (1977) suggested that two or three dominant genes are responsible for β-glucan levels. An additive genetic system of three to five effective factors that control β-glucan content in barley was described by Powell et al. (1989), but the chromosome location was not known. More recently, Han et al. (1995) reported three QTLs located on chromosome 2(2H) and one QTL located on chromosome 5(1H) that accounted for 34% of the total β-glucan in a Steptoe/Morex mapping population.

It has been pointed out that genetic factors are largely responsible for β-glucan levels in barley, but levels are also influenced by environmental growing conditions. MacGregor and Fincher (1993) reported that β-glucan contents of barley grain range from 2.0 to 11.0%, but usually fall between 4.0 and 7.0%. Changing the amylose to amylopectin ratio in barley starch, with either the *wax* gene or the *amo1* gene, has a significant effect on increasing β-glucan content. Perhaps the first report of increased levels of β-glucan in waxy barley was that of Ullrich et al. (1986). Xue et al. (1997) reported increased β-glucan levels in waxy hulless and covered isolines of Compana and Betzes barleys; 6.1 versus 4.85% ($n = 18$ per mean) in waxy and nonwaxy types, respectively. Ajithkumar et al. (2005) reported a range from 5.6 to 5.75% β-glucans in four waxy lines developed at Svalöf Weibull AB (Svalöv, Sweden) compared to 4.7% in a nonwaxy barley cv. Golf. In a review, Fastnaught (2001) presented data from 16 separate studies showing that β-glucan levels were increased by 32 to 41% in waxy barley compared to nonwaxy barley. Much higher levels of β-glucans have been reported in two unrelated waxy hulless barleys. Azhul is a six-rowed cultivar that most consistently contains the highest levels of β-glucans (10 to 11%) in an otherwise normal barley (Bengtsson et al. 1990; Bhatty 1992; Danielson et al. 1996). Prowashonupana, a chemically induced mutant barley (Eslick 1979, 1981), contains higher levels (15 to 16%) of β-glucans (Hofer 1985), but this is due in part to a shrunken endosperm characteristic.

Similar increases of β-glucans were reported in high-amylose cultivars (Xue 1992; Swanston et al. 1997; Ajithkumar et al. 2005), although the numbers of observations were not as large as that with the waxy types. According to Ullrich (2002), the waxy and high-amylose genes interact with each other when the two traits are introduced into the same genotype. Accordingly, the β-glucan and starch levels are increased and decreased, respectively, exceeding simple additive effects. Stahl et al. (1996) concluded that the association of high levels of β-glucan with the waxy gene could not be broken. Fastnaught (2001) suggested that in waxy cultivars, the genetic blockage of amylose production resulted in glucose that is normally shunted to the starch biosynthetic pathway to be taken up by the β-glucan biosynthetic pathway. In addition to causing an increase in the level in waxy barley, the β-glucan molecular structure is also changed in that there is a higher ratio of β-(1,3)-linked cellotriosyl units to β-(1,3)-linked cellotetraosyl units (Jiang and Vasanthan 2000; Wood et al. 2003).

It is important to note that although the general expectation is an increased level of β-glucan in waxy barley, experimental waxy lines have been developed that contain a genetic factor responsible for reduced β-glucan content (Swanston 1997). These barleys were developed by crossing Chalky Glenn, a barley cultivar having thin endosperm cell walls (a characteristic associated with low β-glucan levels), with a waxy cultivar. In an earlier study, Aastrup (1983) obtained two low-β-glucan barley lines by treating Minerva with a mutagenic chemical. Endosperm cell walls in the mutant lines were much thinner than those of the parent barley, 3.0 μm versus 6.5 μm. β-Glucan levels in the two mutant barleys ranged from 2.0 to 2.7% compared to 4.9 to 5.9% in the parent barley, which had much thicker cell walls. The reports by Aastrup (1983) and Swanston (1997) suggest that there are distinctive genes controlling cell wall thickness that can be manipulated to achieve low β-glucan and potentially higher β-glucan levels.

The single most important genetic factor affecting the NSP content of barley grain is the *Nud* (hulled)/*nud* (hulless) gene, located on chromosome 1(7H). This characteristic is simply inherited and is probably one of the most easily discernable genetically controlled traits in barley. This factor controls the presence or absence of the hull on the caryopsis (Nilan 1964). In the presence of the *Nud* gene, the lemma and palea adher tightly to the caryopsis, remaining intact during harvesting, whereas in the presence of the *nud* trait, they are loosely attached and most are removed at harvest. The cementing substance causing hull adherence in hulled barley is secreted by the caryopsis and appears about 16 days after pollination. Contact between the hulls and the pericarp is not established completely until grain filling is almost complete, which is about 10 days before a kernel reaches its maximum size (Harlan 1920). The hulls then adhere to the pericarp epidermis except over the embryo (Gaines et al. 1985). Major differences in the extent of looseness of the hull attachment in hulless types have been noted in certain genotype backgrounds (R. T. Ramage, personal communication). Barley breeding techniques have been used to produce nearly isogenic isolines of hulled and hulless barley in several genotype backgrounds (Hockett 1981). With the removal of the hull, components of the aleurone, endosperm, and embryo increase in relative proportions due to the removal of cellulose, lignin, and silica in the hull (McGuire and Hockett 1981). For example, starch and β-glucan contents were increased by 9.7 and 6.8%, respectively, while TDF was reduced by 28.8% in nonwaxy hulless isotypes of Compana and Betzes barleys (Xue et al. 1997). Low- and high-β-glucan barleys were studied by Zheng et al. (2000). Low-β-glucan hulless barleys had a greater concentration of β-glucans in the subaleurone layer, with slightly declining amounts toward the center of the kernel. In high-β-glucan hulless cultivars, the β-glucans were evenly distributed throughout the entire endosperm.

The major effects of the *nud* and *wax* genes on carbohydrate composition are shown in Table 4.2. The data shown are typical of most hulled, hulless, waxy and nonwaxy barley, but large variations from these values have been reported (Hofer 1985; Danielson et al. 1996; Zheng et al. 2000).

Proteins

The level of protein in barley is highly variable, ranging from 7 to 25% according to a large USDA study involving over 10,000 genotypes (Ullrich 2002), although the level in typical barley is more commonly reported between 9 and 13% (Duffus and Cochrane 1993; Newman and Newman 2005). In a survey that included 3817 samples of malting barley from western Canada, LaBerge and Tipples (1989) reported that the mean protein content was 12.2%. As a general rule, barley grown under dryland conditions in the western United States will be 2 to 3% higher in protein than barley grown under more moist conditions in the Pacific coastal area or the upper Midwestern region.

Proteins are large-molecular-weight nitrogenous compounds whose basic components are L-amino acids, grouped into peptides. Proteins occur in many forms in the barley kernel, being responsible for structural functions, metabolic activity, and providing nitrogen for the developing embryo at germination. Carbohydrates and lipids are often chemically bound to proteins in barley, and as such are classified as glycoproteins and lipoproteins, respectively, and in some instances are called glycolipoproteins if they contain both protein and lipids. Also, some barley proteins contain minerals such as calcium, phosphorus, iron, and copper. By virtue of the presence of nitrogen at a somewhat constant amount, in barley as in other foods, the protein level is generally measured by determining the nitrogen content by one or more methods. However, not all nitrogen in cereal grains is from true protein, as there are also varying amounts of nonprotein nitrogen (NPN) compounds in the kernel. In recognition of this, protein measurements are often referred to as *crude protein*. True protein (nitrogen from L-amino acids) in barley amounts to 80 to 85% of the crude protein value determined in standard laboratory procedures (Briggs 1978). As with other cereals, the protein level in barley is, on average, relatively low, and the quality of the protein is not exceptionally high compared to that of meat, poultry, or dairy products. *Quality* in protein refers to the amounts and balance between the essential and nonessential amino acids. Essential amino acids are those that cannot be synthesized by an animal's metabolic system. Depending on the species of animal, between 8 and 11 amino acids are classified as essential. Nonessential amino acids are not required to be preformed in the diet, as they can be synthesized by the animal's metabolic system from various carbon skeletons and amine nitrogen. A total of 18 L-amino acids have been identified in barley protein, with lysine and threonine being the first and second most limiting in terms of meeting animal growth requirements. Methionine and tryptophan are the third and fourth most limiting essential amino acids, respectively. Average amino acid composition of typical hulled and hulless barley at three levels of protein is shown in Table 4.3. As with most nutrients in barley other than fiber, removal of the hull has the effect of increasing the protein and amino acid levels in the remainder of the kernel.

T. B. Osborne (1924) pioneered the study of cereal proteins by classifying them into four groups—albumins, globulins, prolamins, and glutelins—on the basis of solubility in a series of solvents. Albumins and globulins are soluble

TABLE 4.3 Amino Acid Composition (g/kg) of Hulled and Hulless Barley at Three Levels of Protein, Dry Matter Basis

Protein and Amino Acids	Level 1[a]		Level 2[b]		Level 3[c]	
	Hulled	Hulless	Hulled	Hulless	Hulled	Hulless
Protein ($n \times 6.25$)	15.7	16.0	13.2	14.0	11.2	11.7
Amino acid						
Alanine	—[d]	—[d]	0.44	0.47	0.35	0.38
Arginine	0.71	0.71	0.60	0.64	0.45	0.50
Aspartic acid	—[d]	—[d]	0.71	0.75	0.55	0.57
Cystine	0.35	0.35	0.28	0.31	0.20	0.20
Glutamic acid	—[d]	—[d]	2.98	3.24	2.28	2.44
Glycine	—[d]	—[d]	0.42	0.44	0.32	0.34
Histidine	0.27	0.35	0.26	0.28	0.20	0.22
Isoleucine	0.54	0.53	0.43	0.46	0.34	0.37
Leucine	1.11	1.16	0.79	0.84	0.67	0.71
Lysine	0.38	0.41	0.41	0.41	0.31	0.34
Methionine	0.39	0.40	0.20	0.28	0.15	0.17
Phenylalanine	0.83	0.86	0.68	0.73	0.51	0.53
Proline	—[d]	—[d]	1.32	1.43	0.96	0.98
Serine	—[d]	—[d]	0.54	0.57	0.41	0.45
Threonine	0.56	0.55	0.42	0.45	0.38	0.37
Tryptophan	—[d]	—[d]	0.22	0.23	0.14	0.15
Tyrosine	0.54	0.55	0.37	0.42	0.32	0.33
Valine	0.75	0.76	0.59	0.63	0.46	0.46

Source: Newman and Newman (2005).
[a]Bhatty et al. 1979, $n = 2$.
[b]Montana Agriculture Experiment Station, unpublished data, $n = 8$.
[c]Truscott et al. (1988), $n = 6$.
[d]Not reported.

in water and dilute salt solutions, respectively, and prolamins are soluble in an aqueous alcohol solution. Glutelins are the remaining proteins not extracted by the other solvents. Classical extraction of glutelins may be made with dilute acid or alkali, but it is more usual at the present time to extract them with a detergent (sodium dodecyl sulfate) or a chaotropic agent (urea) in the presence of a reducing agent (Shewry 1993). The crude solubility fractions of protein in barley and other cereals as determined by Osborne have been useful in many studies; however, these classical fractions are highly heterogeneous and can vary with genotype and with agronomic practice (Eppendorfer 1968; El-Negoumy et al. 1979; Kirkman et al. 1982; Miflin et al. 1983; Shewry and Darlington 2002). Because of the heterogeneous nature of barley proteins, they are best considered as two groups defined according to biological function, storage or nonstorage, rather than as the four classical Osborne fractions (Shewry 1993).

Storage proteins include the major alcohol-soluble prolamins, along with glutelins and minor amounts of globulins. Prolamins, called hordeins in barley, are the major storage proteins in the endosperm, accounting for 35 to 50% of the total nitrogen (Kirkman et al. 1982). Osborne coined the name *prolamins* for this distinct group of proteins, to reflect the high content of proline, a nonessential amino acid and amide nitrogen. Barley hordein is a complex mixture of polypeptides that has been divided on the basis of amino acid composition into three groups: sulfur rich (B + γ), sulfur poor (C), and high-molecular-weight prolamins (D). Amino acid analyses of these groups revealed a composition of 30 mol % or more of the nonessential amino acids glutamic acid and glutamine, and less than 1 mol % of the essential amino acid lysine. Glutelins were initially thought to be a distinct group of proteins; however, it has been determined that many of the component proteins of glutelins are structurally related to prolamins (Shewry and Tatham 1990). Although not soluble in aqueous alcohol, these components may be considered as prolamins since the individual reduced subunit peptides are alcohol soluble and rich in prolamine and glutamine. Glutelins contain less lysine than albumins and globulins, and they are somewhat higher in lysine than true hordein (prolamin) in barley (Shewry et al. 1984; Shewry and Tatham 1987).

The nonstorage proteins, albumins and globulins, are structural components of cell walls and metabolic proteins such as enzymes. Some of the nonstorage proteins may accumulate in large enough quantities as to have a secondary storage role. Nonstorage proteins are found primarily in the aleurone and embryo, accounting for 15 to 30% of the total grain nitrogen. Lysine comprises 5 to 7% of the albumin–globulin fraction, compared to less than 1% in hordeins of most barley, and is intermediate between these two levels in glutelins. Threonine is also found in greater concentration in the albumins and globulins. Thus, as a result of increased lysine and threonine, the nonstorage proteins are more nutritionally balanced than the storage proteins (Munck 1972; Brandt 1976; El-Negoumy et al. 1977; Kirkman et al. 1982; Shewry et al. 1984).

Protein and amino acid levels in barley are influenced strongly by both genetic and environmental growing conditions, although the latter has the greater effect. Available soil nitrogen is a major environmental factor influencing protein levels and amino acid levels. Even though total amino acid levels are increased with elevated protein synthesis in the presence of ample soil nitrogen and moisture, they are not increased proportionally to one another. Rather, individual amino acids that are produced are representative of the composition of storage proteins (hordeins), which are increased selectively by soil nitrogen. Therefore, high protein levels in barley increased by nitrogen fertilizer generally mean high levels of the nonessential glutamic acid, proline, and glutamine and lower levels of the essential amino acids, particularly lysine and threonine. The negative correlation between protein and lysine levels in most barley is due to amino acid composition differences in the four major protein fractions and the overwhelming prevalence of low-lysine hordein in most high-protein barley.

Genetic control of protein level in barley is quantitatively inherited and is considered to be extremely complex. Only limited genetic information about total grain protein content was determined prior to the development of *quantitative trait locus* (QTL) analysis (Ullrich 2002). Shewry (1993) presented a detailed summary of the genetic control of hordein synthesis in barley which indicated that all of the hordein proteins are encoded by structural genes on chromosome 5(1H). Through the use of QTL analysis, it is now recognized that there are regions on all seven chromosomes associated with the control of barley kernel protein content (Zale et al. 2000; Ullrich 2002).

Prior to the advent of QTL analysis, the lysine level in cereal grains was studied intensively in numerous laboratories during the late 1950s and early 1960s. At that time, the research was driven by widespread famine and protein deficiency in many developing countries. The U.S. Agency for International Development provided a major part of the funding for this research, which was conducted at international research centers, including CIMMYT in Mexico and ICARDA in Syria, and at various state agricultural research stations in the United States. Research centers in Denmark and Sweden also made major contributions to the development of high-lysine barley. Since lysine is the most limiting essential amino acid in barley and other cereal grains, this amino acid received the greatest amount of research attention. Following the discovery and report of high-lysine maize by Mertz et al. (1964), high-lysine cultivars were discovered in grain sorghum (Mohan and Axtell 1975) and barley (Munck et al. 1970). A high-lysine cereal grain is defined as one having $\geq 4.0\%$ lysine in the protein. The concentration is often expressed as grams of lysine per 16 g nitrogen. Barley containing 11 to 13% protein will contain 0.40 to 0.45% lysine (w/w %) and about 3.50% lysine when expressed as grams per 16 g N, whereas a high-lysine barley having a comparable protein level may contain up to 0.65% lysine (w/w %).

Lysine in barley is a simply inherited characteristic (Ullrich 2002). Following the discovery of the naturally occurring high-protein, high-lysine cultivar Hiproly, numerous high-lysine barley mutants were induced by chemical and radiation treatment, and most have been characterized biochemically and genetically (Ullrich 2002). Possibly the most studied of the high-lysine types were Hiproly (about 4.50% lysine grams per 16 g N) (Munck et al. 1970) and a mutant barley induced by radiation, Bomi Risø 1508 (about 5.50% lysine grams per 16 g N), produced from the Danish cultivar Bomi (Ingversen et al. 1973) The 1508 barley is considered a hordein mutant, as the change in lysine is due to a large reduction in hordein (Shewry 1993), whereas the lysine increase in Hiproly is due to increased amounts of enzymes such as β-amylase that are high in lysine (Hejgaard and Boisen 1980). Accordingly, the genes producing the high-lysine effect, *lys1* and *lys3*, are both located on chromosome 7(5H) (Karlsson 1972; Jensen 1979). At least 10 high-lysine mutant genes have been reported in barley; however, the biochemical and morphological alterations on the kernels resulted in significant reduction in agronomic production (Ullrich 2002). This problem was 90 to 95% overcome by crossing Bomi Risø 1508 (*lys3a*) with two Danish malting varieties, Triumph and Nordal. A selection from this breeding program produced a cultivar designated CA 700202, which contained about 6.50% lysine grams per 16 g N. In yield trials CA 700202 equaled the performance of the original check variety but failed to equal the top-yielding cultivars by the time of its development. The most complete review of the high-lysine barley breeding saga and the genetics involved was written by Lars Munck (1992).

Lipids

In a review on barley lipids, Morrison (1993a) stated that lipids have received much less attention by researchers than other nutritional components, such as carbohydrates and protein. Even so, this author presented a wealth of information on barley lipids in his review, including classification, composition, distribution in the kernel, varietal differences, and relationship with other components. Lipids are located throughout the barley kernel and are classified into two basic fractions: nonstarch lipids and starch lipids. Nonstarch lipids include all lipids other than those inside starch granules. Lipids are also classified as polar or nonpolar, a characteristic that affects solubility and depends on molecular structure. Additionally, the term *free lipid* has been used to describe those lipids that can be extracted with nonpolar solvents, and *bound lipids*, those that are extracted with polar solvents. The proportion of free and bound lipids varies between tissues, and Morrison (1993a) cautions that it should never be assumed that free lipids are the same as total lipids.

Morrison (1993a) also cautions that the choice of analytical procedures for measuring lipid content is critical in accurate determination of the true total lipid content values and/or accurate separation of the various types and classes of lipids. Solvent extraction followed by gravimetric measurement is the normal approach to measuring total lipid content in cereal grains; therefore, effective extraction depends on proper sample preparation (grinding), choice of solvent, and acid hydrolysis prior to extraction. Åman et al. (1985) reported that the total lipid content of 115 barley samples measured after acid hydrolysis ranged from 2.1 to 3.7% (weighted mean = 3.0%). In a later study in this laboratory, Åman and Newman (1986) reported somewhat higher values in 20 high-lysine (2.6 to 5.9%) and three high-sugar barleys (4.4 to 7.3%). Lower mean levels of total lipids, in the range 2.0 to 2.5%, have been reported in older literature and composition text tables; however, it may be assumed that these values were obtained without acid hydrolysis prior to extraction (Morrison 1993a). Parsons and Price (1974) reported total lipid levels ranging from 3.1 to 4.6% in 18 varieties that were extracted with polar solvents. These varying reports emphasize the necessity of standardized analytical techniques and being certain of the technique utilized when reporting compositional data.

Lipid is concentrated in the embryo, and although the embryo represents only about 3% of the kernel by weight, it furnishes about 18% of the total lipid in barley. The endosperm contains about 3% lipid, but by virtue of size, furnishes about 77% of the total lipid, Most of the endosperm lipid is found in the aleurone layer. The remaining 5% of the total kernel lipid is found in the hull (Price and Parsons 1975, 1979; Briggs 1978). According to Briggs (1978), the aleurone contains about 67% of the neutral lipids of the grain total. Nonpolar lipids represent about 75% of the total lipids in the whole grain, with the remaining 25% somewhat evenly divided between glycolipids and phospholipids (polar lipids). In general, this same ratio of nonpolar and polar lipids is representative of lipids in bran, endosperm, aleurone, embryo, and hull.

Nonstarch lipids are stored in oil droplets called *spherosomes* surrounded by a membrane. These membranes contain polar lipids, principally phospholipids (Morrison 1993b). The principal nonpolar lipid in the droplets is triacylglycerol, with the remainder being comprised of steryl esters, diacylglycerol, monoacylglycerol, and free fatty acids. The major fatty acids in barley triacylglycerol are palmitic acid (16:0), oleic acid (18:1), linoleic acid (18:2), and linolenic acid (18:3), comprising approximately 23, 13, 56, and 8%, respectively of the total. The saturated fatty acid stearic acid (18:0) makes up less than 1% of the total (Anness 1984; Morrison 1993a). Morrison (1993a) also noted that the fatty acids in barley are similar to those in wheat except that barley tends to have more linolenic acid than wheat. Larger barley kernels tend to have higher values for palmitic acid and lower values for linoleic and linolenic acids (De Man and Bruyneel 1987). Variety, growing environments, and nitrogen fertilization were reported to have only small effects on fatty acid composition (De Man 1985; De Man and Dondeyne 1985).

Starch lipids are located inside starch granules and in quantities generally proportional to amylose content. The true starch lipids are almost exclusively phospholipids (Morrison et al. 1984; Morrison 1988, 1993a); however, Acker and Becker (1971) found a small amount (4.4%) of free fatty acids, in a sample of barley starch lipid along with phospholipids. These authors reported that palmitic acid was the predominant acid in the free fatty acids, followed by linoleic and oleic acids in that order. Tester and Morrison (1992) reported differences in the fatty composition of starch lipids isolated from waxy and nonwaxy barleys. Palmitic acid and linoleic acid contents varied inversely in total starch lipids of waxy and nonwaxy mature grains. Total starch lipids in waxy barley contained about 30% more palmitic acid and about 30% less linolenic acid than did nonwaxy barley.

Genetic selection and breeding for increased total lipids in barley kernels have met with only limited success. Price and Parsons (1974) reported finding a barley cultivar (CI 12116) grown in South Dakota that contained 4.6% total lipid, but subsequent samples of this barley grown at two locations in Canada contained normal (2 to 3%) lipid levels (Fedak and de la Roche 1977; Bhatty and Rossnagel 1979). The high-lysine mutant Risø 1508 was reported to contain higher levels of total lipids (Bhatty and Rossnagel 1979; Newman et al. 1990; Munck 1992); however, it was suggested by Morrison (1993a) that the increased levels were due to the severely shrunken endosperm of the mutant barley. However, it is worthy of note that a high-lysine cultivar (CA 700202) developed from Risø 1508 in Denmark (Bang-Olsen et al. 1987) has plumper kernels with a higher starch content than the mutant parent, yet maintains the higher level of total lipids (Newman et al. 1990). Another shrunken endosperm mutant that has waxy starch is reported to contain about 7.0% total lipid (Hofer 1985; Åman and Newman 1986). Some reports suggest that the genetic factors controlling waxy and high-amylose starch in barley may be linked to genetic factors that increase levels of total lipids. Xue et al. (1997) reported a 25% increase in total extractable lipids in waxy compared to nonwaxy isotypes in Compana and Betzes isotypes ($n = 18$). Glacier Ac38, a high-amylose selection from the cultivar Glacier, is reported to contain over 30% more total lipid than is contained by Glacier (2.3 vs. 3.0%). Hulless high-amylose Glacier developed in a backcrossing selection study at Montana State University was reported to contain 2.9% total lipid, thus retaining a major portion of the increased lipid in the original high-amylose cultivar, Glacier Ac38 (Xue 1992). Although there appears to be a linkage in increased lipids with high-lysine and starch type, caution should be exercised in interpreting these data without more complete testing and evaluation of larger samples grown in diverse environments.

An important group of nutrients, associated with barley lipids due to their solubility in lipid solvents, are the tocopherols and tocotrienols. This group, collectively called *tocols* or vitamin E complex, is discussed in the following section.

Vitamins

Vitamins are organic compounds required in small amounts for the maintenance of normal biochemical and physiological functions of a mammalian system. These compounds are essential components of a diet, and if absent for a period of time, deficiency symptoms are manifested in various ways. Vitamins are referred to as *accessory food factors*, and as such, they are considered essential nutrients in that they cannot be synthesized in amounts sufficient to meet the needs of a system. The vitamin theory was initiated in the early part of the twentieth century (Funk 1911, cited by Oser 1965) and the term *vitamine* was coined by Funk. The terminal "e" was dropped from the spelling, as subsequent research showed that compounds fitting the definition of accessory food factors were not all amines. The term *vitamin* has been retained in its generic rather than its chemical sense and presently is used as a name for accessory food factors that are neither amino acids nor inorganic elements.

The vitamins were originally named according to their function, their location, the order in which they were discovered, and in combinations. Over time, at least 68 different accessory food factors were proposed, but only 15 compounds are officially recognized as true vitamins at the present time (Berdanier 2002). The vitamins are divided into classes based on solubility in lipid solvents or water. The lipid-soluble vitamins are A, D, E, and K; the water-soluble vitamins are thiamine (B_1), riboflavin (B_2) nicotinic acid (B_3), pyridoxine (B_6), cobalamine (B_{12}), biotin, pantothenic acid, folic acid, ascorbic acid (vitamin C), choline, and myoinositol. Other than vitamin C and myoinositol, the water-soluble vitamins are often referred to as the B-complex vitamins. Since some synthesis of choline occurs in animals, it does not strictly adhere to the definition of an accessory food factor and is thus not always considered a vitamin.

Cereal grains have long been noted as good sources for certain vitamins, especially some of the B-complex vitamins. The barley caryopsis contains all of the vitamins and choline with the exception of vitamins A, D, K, $B_{12,}$ and C. Reported analytical values for many of the vitamins in barley often disagree widely, the differences being either real or due to analytical techniques (Briggs 1978). Recently, there has been a surge in investigative activity concerning vitamin E in cereals, due to reported effects of some of the vitamin E components on human health. Vitamin E is a complex of eight isomers, four tocopherols and four tocotrienols, collectively called *tocols*. Of the major cereals, barley contains the highest amount of total tocols, according to Kerckhoffs et al. (2002). The tocols are composed of a chromanol ring with an attached phytyl C16 side chain. Tocotrienols (T3) differ from tocopherols (T) in that their phytyl side chain contains three double bonds at carbons 3, 7, and 11. The four isomers in T and T3 differ in the number and positions of methyl groups on the chromanol ring; α-(5,7,8-trimethyl), β-(5,8-dimethyl), γ-(7,8-dimethyl), and δ-(8-methyl) (Hunt and Groff 1990). For many years the biological activity of vitamin E has been ascribed primarily to α-tocopherol; however, McLaughlin and Weibrauch (1979) calculated a vitamin E equivalent value utilizing a sum of α-tocopherol and percentages of α-tocotrienol and the other isomers.

The tocols are associated with lipid components in aleurone, endosperm, and embryo tissue, and concentrations are positively correlated with oil content. α-Tocopherol and α-tocotrienol concentrations in barley oil were reported to be 24 and 17 times greater, respectively, that those in corn oil (Wang et al. 1993). Peterson (1994) showed that hull, endosperm, and embryo of Morex barley contained 5.4, 0.6, and 93.7%, respectively, of the total tocopherol and 25.9, 45.7, and 28.5%, respectively, of the total tocotrienols. Thus, the majority of tocopherols are found in the embryo, whereas tocotrienols are more evenly dispersed throughout the kernel.

Cavallero et al. (2004) confirmed earlier reports of Peterson and Qureshi (1993) and Wang et al. (1993) for the total amount and relative abundance of T and T3 isomers in barley. Cavallero et al. (2004) indicated that both genotype and environment influenced total tocol content. A recent investigation indicated a wide range of tocol concentration in 13 different barley cultivars produced in three growing seasons under two cropping systems (Ehrenbergerová et al. 2006). Differences in tocol concentrations were associated not only with cultivar but also with climate (year) and soil fertilization practice. In this study, total tocols (T + T3) in the kernel ranged from 46.7 to 67.6 m/kg, with an average of 53.8 mg/kg. Tocotrienols comprised 77% of the total tocols, with 23% coming from tocopherols. Of the eight isomers, α-T3 was the major compound, comprising 73% of the total tocols in these barleys. These authors reported higher levels of all tocols in hulless and waxy barley types. Wabet, a waxy hulled cultivar used in a reciprocal crossing program, resulted in a higher content of tocols and the isomers of α-T and α-T3 in its progeny than that in the other waxy varieties, confirming the findings of Vaculová et al. (2001), cited by Ehrenbergerová et al. (2006). In another recent study, Moreau et al. (2007) evaluated the tocol content of two hulled and two hulless barley cultivars. One of the hulless types was waxy and was grown in a different environment and year than the other three barleys. The concentrations of all T and T3 isomers reported in this study were higher than reported by Wang et al. (1993), Peterson (1994), and Ehrenbergerová et al. (2006). The greatest differences were in α-T and α-T3 concentrations and the sum of the eight isomers, which was 2.0 to 2.6 times greater than in the earlier reports. In contrast to the findings of Ehrenbergerová (2006), Cavallero et al. (2004) and Moreau et al. (2007) reported higher levels of total tocols in hulled barley although they also reported that hulless types had higher levels of α-T3. The results of these four studies indicate variability in concentrations of the various tocol isomers among different barleys. The differences may be explained by the different laboratories, genotypes, and growing environments. The one consistent finding in all four studies was the greater concentration of α-T3 than of the other T and T3 isomers (Table 4.4).

There is much less recent information on the amounts of the other vitamins in barley grain. Whole-grain cereals in general are good sources of the B-complex vitamins. Barley contains a range of the B vitamins that act as cofactors or prosthetic groups for various metabolic reactions in animal and microbial metabolism. A summary of the B-complex vitamin concentrations in barley grain is presented in Table 4.5. The data, which are from different laboratories spanning several years, are consistent for some vitamins but vary widely for others. It was suggested by Briggs (1978) that variation in the concentrations of vitamins reported in barley may be attributed to differences in analytical procedures and techniques used in different laboratories. Nicotinic acid is the most abundant of the

TABLE 4.4 Tocol Content (mg/kg) of Barley, Dry Matter Basis

η	Tocopherols (T)				Tocotrienols (T3)				T+T3	T3(%)
	α	β	γ	δ	α	β	γ	δ		
7[a]	10.9	0.6	11.0	0.3	32.6	—	6.9	0.7	63.0	63.8
2[b]	8.1	0.4	10.0	0.4	43.3	—	7.7	1.0	70.9	73.3
3[c]	7.6	0.5	0.4	0.2	26.0	5.3	4.2	0.7	44.9	80.6
4[d]	5.1	0.5	3.7	0.3	16.6	3.3	8.1	1.2	38.8	75.3
6[e]	9.3	0.4	2.1	0.2	30.8	4.3	6.9	0.8	54.8	78.1
13[f]	6.8	4.3[h]	—	1.0	29.6	11.4[h]	—	0.6	53.7	77.5
4[g]	17.9	1.5	10.3	2.6	58.7	8.9	13.6	2.3	115.8	72.1
Weighted means	8.7	0.7	6.4	0.8	32.3	5.3	7.9	0.9	63.0	73.7

[a] Wang (1992).
[b] Wang et al. (1993).
[c] Peterson (1994).
[d] Andersson et al. (2003).
[e] Cavallero et al. (2004).
[f] Ehrenbergerova et al. (2006).
[g] Moreau et al. (2007).
[h] β and γ isomers were combined in this report and not included in weighted means.

TABLE 4.5   B-Complex Vitamins (mg/kg) in Barley, Dry
Matter Basis

	n	Means	Range
Thiamine	4	5.2	4.0–6.5
Riboflavin	5	1.8	1.2–2.9
Nicotinic acid	4	63.3	46.0–80.0
Pantothenic acid	4	5.1	2.8–8.0
Biotin	2	0.14	0.13–0.15
Folic acid	3	0.43	0.2–0.7
Pyridoxine	3	3.5	3.0–4.4
Choline	2	1290	920–2200

Sources: Hopkins et al. (1948); Briggs (1978); National Research Council (1984); Kent and Evers (1994); Köksel et al. (1999); USDA (2005).

B-complex vitamins in barley, although it is 85 to 90% biologically unavailable. Extrapolation from research with wheat bran indicates that nicotinic acid is incorporated in a number of polysaccharide and glycopeptide macromolecules that do not release the vitamin unless treated with alkali or acid (Mason et al. 1973; Carter and Carpenter 1982). Barley contains significant amounts of tryptophan, from which nicotinic acid may be synthesized by animals, although approximately 50 mol of tryptophan is required to produce 1 mol of nicotinic acid.

Variation in the content of B-complex vitamins from one cereal grain to another is quite small except for nicotinic acid, which is much higher in barley, wheat, sorghum, and rice than oats, rye, maize, and millet. Barley contains the highest level of nicotinic acid of all the cereals. Extrapolating from wheat data, the B-complex vitamins are concentrated in various parts of the barley kernel. Thiamine is concentrated primarily in the scutellum (62%) and aleurone layer (32%). Almost equal amounts of riboflavin found in the aleurone layer (37%) and endosperm (32%), with smaller amounts occurring in the embryonic axis (12%) and scutellum (14%). Nicotinic acid is concentrated in the aleurone layer (61%), with smaller amounts in the pericarp (12%). Pyridoxine is also concentrated in the aleurone layer (61%), with smaller amounts in the pericarp (12%) and scutellum (12%). Pantothenic acid is concentrated in the aleurone layer (41%) and endosperm (43%) (Kent and Evers 1994).

Minerals

The gross mineral matter of barley is the "ash" remaining after burning a sample until it is free of carbon. Thus, minerals are often referred to as "inorganic" nutrients. The ash content of typical barley ranges from 2.0 to 3.0%, with hulless types being on the lower end and hulled types on the upper end of the range. Barley hulls contain around 6.0% ash (Kent and Evers 1994), which is 60 to 70% nutritionally inert silicon occurring primarily in the outer lemma (Liu and Pomeranz 1975; Pomeranz 1982). At least 14 mineral elements shown to be nutritionally important occur in varying amounts in whole-grain barley. The minerals are divided into two groups, macro and micro elements, on the basis of their concentration in foods. The macro elements found in barley are calcium, phosphorus, potassium, magnesium, sodium, chlorine, and sulfur, and the micro elements are cobalt, copper, iron, iodine, manganese, selenium, and zinc. Silicon is a nonnutrient mineral associated with lignin and cellulose in the hull. Other elements, such as chromium, nickel, and aluminum, have been identified in barley in very small amounts, but their nutritional significance is questionable or unknown. Mineral element composition data of hulled barley are presented in Table 4.6. Despite the fact that mineral composition in barley may vary considerably (Owen et al. 1977; Briggs 1978), these data taken from different reports are remarkably consistent. The greatest variation in the data between reports were in iron, 36 to 85 mg/kg, and phosphorus, 0.26 to 0.44 g/kg, which given the small magnitude of total amounts is probably of little nutritional significance. A recent study involving three barley varieties from Turkey reported values nearly identical to those shown in Table 4.6 for calcium and magnesium, copper, iron, and zinc, but the Turkish barley contained four times more manganese (Köksel et al. 1999).

Mineral elements are distributed throughout the seed, but the greatest concentrations are found in the embryo, pericarp, and aleurone (Liu et al. 1974; Weaver et al. 1981; Marconi et al. 2000). Liu and Pomeranz (1975) determined with x-ray analysis that potassium had the highest concentration in the aleurone but

TABLE 4.6   Mineral Composition of Barley, Macro and Micro, Dry Matter Basis

		Hulled Barley	
Item	n	Mean	Range
Macro (g/100 g)			
Calcium	7	0.05	0.03–0.06
Phosphorus	7	0.35	0.26–0.44
Potassium	7	0.47	0.36–0.58
Magnesium	7	0.14	0.10–0.18
Sodium	6	0.05	0.01–0.08
Chloride	4	0.14	0.11–0.18
Sulfur	2	0.20	0.16–0.24
Silicon	3	0.33	0.15–0.42
Micro (mg/kg)			
Copper	7	6.25	2.0–9.0
Iron	8	45.7	36.0–85.0
Manganese	7	27.2	17.0–20.0
Zinc	6	34.4	19.0–35.0
Selenium	4	0.4	0.2–0.5
Cobalt	3	0.07	0.05–0.10

Sources: Morrison (1953); Weaver et al. (1981); National Research Council (1984); Kent and Evers (1994); Sugiura et al. (1999); Li et al. (2001a, 2001b); USDA (2005).

was distributed in all portions of the kernel. Calcium was highest in the aleurone, with smaller amounts in the lemma and endosperm cell walls. Magnesium and phosphorus were most concentrated in the aleurone, with lesser amounts in the endosperm. Sulfur and chlorine were confined to cell walls of the aleurone and endosperm.

Potassium and phosphorus are the most abundant minerals in barley, and the latter is possibly the most important in nutritional terms. Phytic acid (myoinositol) is the most abundant form of phosphorus in barley, representing 65 to 75% of total kernel phosphorus (Raboy 1990). Phytic acid is localized in barley embryo and aleurone tissues (O'Dell et al. 1972), with 80% found in the embryo and 20% in the aleurone. Phosphorus in phytic acid is considered bound and unavailable for utilization by monogastric animals, as they do not possess the enzyme phytase in the digestive tract. However, in preparing foods with alkali treatment, phosphorus in phytic acid becomes biologically available. A major disadvantage of unaltered phytic acid is that divalent cations such as calcium, copper, and zinc are irreversibly chelated with phosphorus in the molecule, making them nutritionally unavailable.

Very little genetic research has been conducted on the mineral content of barley, with the exception of phosphorus (Ullrich 2002). Mutant barley genotypes that have significantly reduced the level of phytic acid have been developed by U.S. Department of Agriculture scientists (Larson et al. 1998; Raboy and Cook 1999). The low-phytic-acid barley shows a reduction of 50 to 75% in phytic acid phosphorus with no obvious change in total seed phosphorus. Compared with normal barley, phosphorus in the low-phytic-acid barley has been shown to be more available to fish (Sugiura et al. 1999), swine (Veum et al. 2002), and poultry (Li et al. 2001a, b). Two genes responsible for the low-phytic-acid mutations, *lpa1* and *lpa2*, have been mapped to chromosomes 2(2H) and 1(7H), respectively (Larson et al. 1998). Cultivars with homozygous *lpa1* gene have various amounts of phytic acid phosphorus reduction, with a concomitant increase in inorganic phosphorus and no change in total phosphorus. In homozygous *lpa2* cultivar types, kernel phytic acid phosphorus is reduced, accompanied by an increase in inorganic phosphorus. The genetics, breeding, and nutritional implications of low-phytic-acid cereals, including barley, has been thoroughly reviewed (Larson et al. 1998; Raboy et al. 2001; Dorsch et al. 2003). Although mineral element concentrations are known to be affected by variety (genetics), as in the case of the low-phytic-acid mutant, growing season, soil type, moisture, and fertility, will also have measurable effects (Kleese et al. 1968; Owen et al. 1977).

Phytochemicals

Phytochemicals are compounds occurring in plants, other than the basic recognized nutrients, that are reported to have protective actions against certain types of cancer, cardiovascular disease, and degenerative diseases such as arthritis, often due to their antioxidant activity. These compounds are sometimes referred to as *phytonutrients*, and when extracted and purified, they are often called *nutraceuticals*. Grains in general contain different types of phytochemicals, and in barley the most researched compounds are sterols, tocotrienols, flavonols, and phenolic compounds (Groupy et al. 1999). Much of the antioxidant activity in grains, including barley, comes from insoluble phenolic compounds such as ferulic acid, which are esterified to cell wall polysaccharides (arabinoxlans) and lignin (Bunzel et al. 2004).

The greatest concentrations of phytosterol compounds are located in the outer layers of the kernel (Lampi et al. 2004). These authors reported 797 and 1738 mg/kg in whole hulless barley (cv. Doyce) and pearling fines, respectively. In a follow-up study in the same laboratory using a seed scarifier, Moreau et al. (2007) removed approximately the same amount of pearling fines (about 11 to 15%) from two hulled (cv. Throughbred and cv. Nomini) and two hulless (cv. Doyce and Merlin) barleys. Total kernel phytosterols averaged 818 and 2349 mg/kg (fresh weight), respectively.

Zupfer et al. (1998) determined the concentrations of ferulic acid in 18 cultivars of two- and six-rowed barleys grown at two locations in Minnesota. The ferulic acid concentrations ranged from 365 to 605 mg/kg of dry weight. The concentration of ferulic acid varied significantly among the cultivars. These authors inferred a genetic basis for ferulic acid concentration due to the similar ranking of the cultivars at the two locations. In a study reported by Holtekjølen et al. (2006), ferulic acid was the most abundant phenolic acid, accounting for 52 to 69% of the total, ranging from 403 to 723 mg/kg in 16 different barley cultivars, including hulled, hulless, waxy, and nonwaxy types. Higher levels of ferulic acids were observed in hulled varieties than in the hulless cultivars. These results were similar to those reported by Hernanz et al. (2001) and Moreau et al. (2007).

Holtekjølen et al. (2006) also reported on the concentration of proanthocyanidins (flavonols). The main flavonols found were the catechins, procyanidin B_3, and prodelphinidin B_3. A waxy hulless Canadian barley, CDC Alamo, contained the highest catechin content, almost three times the levels in the cultivars with the lowest levels. Total flavonol content ranged from a low of 325 to a high of 527 mg/kg. The cultivars with the highest total flavonols were a hulled six-rowed Norwegian barley, Thule, followed by CDC Alamo. Ragaee et al. (2006) reported that barley contains more total phenols (as gallic acid equivalent) than hard and soft wheat, but less than millet, rye, and sorghum.

Certain of the phytochemicals found in barley are pigments that produce kernels of many different colors, including white, blue, black, purple, and red; however, only varieties with blue or white kernels are grown extensively, and the white varieties dominate. Black pigmentations of various parts of the barley plant is generally due to the presence of melanins. Red or purple colors are due primarily to anthocyanins (Takahashi 1955). The purple and blue colors in barley kernels are usually the results of a phenolic compound on the surface of the grains, particularly thc anthocyanins. Many shades of the basic colors are possible, due to combinations of anthocyanins and their interactions with other phenolic compounds (Mazza and Miniati 1993). Mazza and Gao (2005) cited reference sources for anthocyanins and other phenolics that can occur in barley, which included six anthocyanins, four proanthocyanins, ten phenolic acids, four phenol glucosides, and one flavonol. The purple color of the kernel is located in the pericarp, while blue is localized in the aleurone layer of the kernel.

Anthocyanin pigmentation in kernels is controlled by two complementary genes, *P* and *C* (Woodward and Thieret 1953). The blue-colored aleurone is governed by the complementary genes *Blx1* and *Blx2* (Mayler and Stanford 1942; *Barley Genetics Newsletter* 1996). Proanthocyanidin-free barleys have been produced through a mutation research program, and the trait is a monofactorial recessive. The proanthocyanidin-free genes have been located on chromosomes 1(7H), 3(3H), and 6(6H) (Jende-Strid 1995). The mutants used most in malt breeding programs are the *ant13* and *ant17* genes, located on chromosomes 6(6H) and 3(3H), respectively (Ullrich 2002). Many of the pigments that occur in barley fall into the phytochemical classification, including anthocyanins, proanthocyanins, flavonols, phenolic acids, and phenol glucosides (Mazza and Gao 2005). For the most part, these compounds occur in the hull and outer layers of the barley caryopsis, as reported by Lampi et al. (2004).

GENETICS AND PHYSICAL CHARACTERISTICS

The most easily visualized physical characteristics of barley are hull type and row type. The former was discussed in the carbohydrate section because of the major effect of the hull on total and classification of dietary fiber and starch content.

Row Type

Row type is a simply inherited characteristic that was described briefly in relation to the anatomy of the developing kernel in Chapter 2. The genetics of two- and six-rowed barleys has been researched, documented, and reviewed extensively (Hockett and Standridge 1976; Gymer 1978; Hockett and Nilan 1985; Ullrich 2002). As with hull type, row type is readily discernible upon visual inspection. Row type is controlled by two genes, *Vrs1* and *Int-c*, which are located on chromosomes 2(2H) and 4(4H), respectively (Nilan 1964; Kleinhofs and Han 2002). In commercially released barleys, two-rowed types have the genotype *Vrs1Vrs1int-cint-c*, and six-rowed types have the genotype *vrs1vrs1Int-cInt-c* (Ullrich 2002). In two-rowed barleys only the central floret is fertile and two lateral florets are sterile, resulting in a single seed at each node, giving the head a flat appearance. In six-rowed barleys all three spikelets are fertile, resulting in differences in the shape of the kernels on the spike, giving the head a round or tubular appearance. In two-rowed barleys all kernels are symmetrical, and the principal variation in size is between the larger kernels in the middle of the spike and the smaller ones at the base and tip of the spike. In six-rowed barleys, approximately one-third of the kernels are symmetrical, as in two-rowed varieties, and the other two-thirds are asymmetrical (twisted). The twist is more pronounced on the attachment end at the rachis and less twisted on the distal end. The twisted kernels are from the lateral rows of the spike, whereas the symmetrical kernels are from the central rows. The twisted kernels are smaller and weigh from 13 to 20% less than the symmetrical kernels. The symmetrical kernels of six-rowed varieties are slightly contracted or pinched in laterally at the attachment end, in contrast with kernels of two-rowed varieties, which are usually broader at this point.

Two-rowed varieties usually have higher numbers of tillers per plant and large, heavier seed than that of six-rowed varieties. On the other hand, six-rowed varieties usually have more seed per spike. Thus, the compensatory effects of yield component lead to similar levels of yield potential. Although two-rowed barleys produce larger kernels than those produced by six-rowed types, in both row types considerable variation exists in kernel shapes, from long and slender to short and plump (Briggs 1978). Two-rowed barleys have been reported to contain slightly more protein than is present in two-rowed types (Western Regional Project-166 1984; Newman and McGuire 1985); however, the amino acid composition in barley is probably more dependent on the total protein in the kernel than the row type (Pomeranz et al. 1973; Newman et al. 1982; Western Regional Project-166 1984; Newman and McGuire 1985). Other factors being equal, such as environmental growing conditions, the nutrient parameters of two- and six-rowed barley are not significantly different.

Kernel Size, Kernel Weight, and Volume Weight

These characteristics of barley are quantitatively inherited and are measures that were developed for evaluating hulled barley. Very little is known of the genetics of these traits, except that it is complex (Ullrich 2002). Removal of the hull either genetically or mechanically alters these measurements so that they are not meaningful except for comparisons within hull or hulless types.

Kernel size or plumpness is accomplished with special sizing screens designed to separate kernels into plump and thin categories. Percentage plumpness is an important criterion for determining the quality grade of malting barley and could become an important measure for quality in food barley. Plump kernels will contain significantly more starch, less protein, and less fiber than will thin kernels. Comparison of plumpness values between hulled and hulless types cannot be justified, due to differences created by removing the hull, but may be useful in comparing barleys within hulless and hulled types. A good plumpness score would be 90 to 95% plump kernels with 10% or less thin kernels.

Thousand kernel weight (TKW) provides a measure of average kernel weight to the nearest 1.0 mg per kernel or the nearest 1.0 g for 1000 kernels. TKW may be expressed on either a dry or an "as is" basis, most often the latter. This measurement complements kernel size distribution (plumpness), as the two are highly and positively related. As with good plumpness value, heavy kernels contain a high percentage of starch. Kernels from two-rowed barleys have higher TKW values than those of six-rowed barley. Data from the 2006 barley nursery at WestBred LLC, Bozeman, Montana (Dale Clark, personal communication) showed TKW mean values (and ranges) of 46 (43 to 50) and 39 (36 to 42) for two rowed ($n = 42$) and six-rowed ($n = 25$) cultivars, respectively. The TKW of hulless barleys is reduced 15 to 20%, due to removal of the hull.

Volume weight, once commonly called *test weight*, is now referred to as *hectoliter weight*. Regardless of the terminology, it is a measure of density and is currently expressed as kilograms per hectoliter (kg/hL). Previously, volume weight was expressed in pounds per bushel. The standard test weight of hulled barley is 62 kg/hL (48 lb per bushel), but may range from 52 to 72 kg/hL. Hulless barley volume weights may be as high as 80 kg/hL (Ullrich 2002).

Kernel Uniformity

The uniformity of kernels at harvest increases the ease of malting and processing for food. Kernels of different sizes vary widely in malting behavior, producing malts with distinctively different properties (Burger and LaBerge 1985). Blends of barley having different-size kernels may also present problems in the pearling process, especially when the barley is only slightly pearled. Edney et al. (2002) reported that the smaller nature of six-rowed kernels required longer pearling times than those for two-rowed kernels. However, these authors stated that there appeared to be no advantage in selecting two- versus six-rowed barley for pearling, other than the difference in time required for processing. Having kernels of relatively the same size and shape may be a desirable characteristic for some processing purposes and may be controlled to some extent by the choice of barley cultivar or row type.

SUMMARY

The major and minor nutrients in barley are presented in this chapter. Starch is the largest component, followed by total dietary fiber and protein, respectively. Starch in most barley is composed of two types, branched chains of glucose (amylopectin) and comparatively straight chains of glucose (amylose) in a 3 : 1 ratio, respectively. Barleys that contain mostly amylopectin starch are called waxy barley, and those that contain more (40 to 70%) amylose starch are called high-amylose barley. Protein levels are shown to vary widely due primarily to cultural practices, but there are genotype effects. Protein quality does not improve with an increased level of protein, which is most often due to increases in the lysine-poor prolamines. High-lysine cultivars have been developed but are not widely grown. The lipid content of most barley is relatively constant, with linoleic acid (18 : 2) being the most abundant fatty acid present in the triacylglycerols. Starch lipid level is highest in high-amylose barley. Barley is rich in the fat-soluble vitamin E and contains varying amounts of the vitamin B complex except vitamin B_{12}. Concentrations of the vitamin E complex, which is composed of eight isomers, have been researched in greater depth and more recently than the B-complex vitamins. Potassium and phosphorus make up the greatest percentage of the ash, with smaller amounts of five other essential macro elements and six essential micro elements. A large percentage of the phosphorus in barley is in the form of phytic acid, which is unavailable to humans and other monogastric animals. Low-phytate barley that contains levels of total phosphorus similar to those in normal barley has been developed. Ferulic acid, one of several phenolic compounds classed as phytonutrients or nutraceuticals, is found in greater concentrations in barley than are other phenolics. Genetic factors that influence composition and physical characteristics of the kernel are discussed, and gene location is indicated where known.

REFERENCES

Aastrup, S. 1979. The effect of rain on the β-glucan content in barley grains. *Carlsberg Res. Commun.* 44:381–393.

Aastrup, S. 1983. Selection and characterization of low β-glucan mutants in barley. *Carlsberg Res. Commun.* 48:307–316.

Acker, L., and Becker, G. 1971. New research on the lipids of cereal starches: II. The lipids of various types of starch and their binding to amylose. *Starch/Stärke* 23:419–424.

Ajithkumar, A., Andersson, R., Christerson, T., and Åman, P. 2005. Amylose and β-glucan content of new waxy barleys. *Starch/Stärke* 57:235–239.

Åman, P., and Newman, C. W. 1986. Chemical composition of some different types of barley grown in Montana, USA. *J. Cereal Sci.* 4:133–141.

Åman, P., Hesselman, K., and Tilly, A.-C. 1985. Variation in the chemical composition of Swedish barleys. *J. Cereal Sci.* 3:73–77.

Anderson, M. A., Cook, J. A., and Stone, B. A. 1978. Enzymatic determination of (1–3) (1–4)-β-glucans in barley grain and other cereals. *J. Inst. Brew.* 84:233–239.

Andersson, A. A. M., Courtin, C. M., Delcour, J. A., Fredriksson, H., Scholfield, J. D., Trogh, I., Tsiami, A. A., and Åman, P. 2003. Milling performance of North European hull-less barleys and characteristics of resultant millstreams. *Cereal Chem.* 80:663–667.

Anness, B. J. 1984. Lipids of barley, malt and adjuncts. *J. Inst. Brew.* 99:315–318.

Aspinall, G. O., and Ferrier, R. J. 1957. The constitution of barley husk hemicellulose. *J. Chem. Soc.* 4188–4194.

Aspinall, G. O., and Ross, K. M. 1963. The degradation of two periodate-oxidised arabinoxylans. *J. Chem. Soc.* 1681–1686.

Bamforth, C. W. 1985. Biochemical approaches to beer quality. *J. Inst. Brew.* 91:154–160.

Bang-Olsen, K., Stilling, B., and Munck, L. 1987. Breeding for yield in high-lysine barley. Pages 865–870 in: *Barley Genetics V: Proc. 5th International Barley Genetics Symposium*, Okayama, Japan.

Banks, W., Geddes, R., Greenwood, C. T., and Jones, I. G. 1972. Physiochemical studies on starches: Part 63. The molecular size and shape of amylopectin. *Starch/Stärke* 24:245–251.

Banks, W., Greenwood, C. T., and Muir, D. D. 1973. Studies on the biosynthesis of starch granules: Part 6. Properties of the starch granules of normal barley and barley with starch in high-amylose content during growth. *Starch/Stärke* 25:225–230.

Barley Genetics Newsletter. 1996. Barley Genome Map. Published online at http://www.wheat.pw.usda.gov/ggpages/bgn/.

Bendelow, V. M. 1975. Determination of non-starch polysaccharides in barley breeding programmes. *J. Inst. Brew.* 81:127–130.

Bengtsson, S., Åman, P., Graham, H., Newman, C. W., and Newman, R. K. 1990. Chemical studies on mixed-linked β-glucans in hulless barley cultivars giving different hypocholesterolemic responses in chickens. *J. Sci. Food Agric.* 52:435–445.

Berdanier, C. D. 2002. Food constituents. Pages 3–95 in: *Handbook of Nutrition and Food*, C. D. Berdanier, ed. CRC Press, Boca Raton, FL.

Bhatty, R. S. 1992. β-Glucan content and viscosities of barleys and their roller-milled flour and bran products. *Cereal Chem.* 69:469–471.

Bhatty, R. S. 1999. The potential of hulless barley. *Cereal Chem.* 76:589–599.

Bhatty, R. S., and Rossnagel, B. B. 1979. Oil content of Risø 1508 barley. *Cereal Chem.* 56:586.

Bhatty, R. S., Christison, G. I., and Rossnagel, B. G. 1979. Energy and protein digestibilities of hulled and hulless barley determined by swine feeding. *Canadian J. Anim. Sci.* 59:585–588.

Bhatty, R. S., and Rossnagel, B. G. 1997. Zero amylose lines of hulless barley. *Cereal Chem.* 74:190–191.

Brandt, A. B. 1976. Endosperm protein formulation during kernel development of wild type and a high-lysine barley mutant. *Cereal Chem.* 53:890–901.

Bunzel, M., Ralph, J., and Steinhart, H. 2004. Phenolic compounds as cross-links of plant derived polysaccharides. *Czech. J. Food Sci.* 22:64–67.

Briggs, D. E. 1978. *Barley*. Chapman & Hall, London.

Burger, W. C., and LaBerge, D. E. 1985. Malting and brewing quality. Pages 367–401 in: *Barley*. Agronomy Monograph 26. D. C. Rasmusson, ed. American Society of Agronomy, Crop Science Society of America, and Soil Science Society of America, Madison, WI.

Carter, E. A., and Carpenter, K. J. 1982. The bioavailability for humans of bound niacin from wheat bran. *Am. J. Clin. Nutr.* 36:855–861.

Cavallero, A., Gianineti, A., Finocchiaro, Delogu, A., and Stanca, A. M. 2004. Tocols in hull-less and hulled barley genotypes grown in contrasting environments. *J. Cereal Sci.* 39:175–180.

Danielson, A. D., Newman, R. K., and Newman, C. W. 1996. Proximate analyses, β-glucan, fiber, and viscosity of select barley milling fractions. *Cereal Res. Commun.* 24:461–468.

DeHaas, B. W., and Goering, K. J. 1972. Chemical structure of barley starches: I. A study of the properties of the amylose and amylopectin from barley starches showing a wide variation in Brabender cooking viscosity curves. *Starch/Stärke* 24:145–149.

De Man, W. 1985. The effect of genotype and environment on the fatty content of barley (*Hordeum vulgare* L.) grains. *Plant Cell Environ.* 8:571–577.

De Man, W., and Bruyneel, P. 1987. Fatty acid content and composition in relation to grain-size in barley. *Phytochemistry* 26:1307–1310.

De Man, W., and Dondeyne, P. 1985. Effect of nitrogen fertilization on protein content, total fatty acid content and composition of barley (*Hordeum vulgare* L.) grains. *J. Sci. Food Agric.* 36:186–190.

Dorsch, J. A., Cook, A., Young, K. A., Anderson, J. M., Bauman, A. T., Volkmann, C. J., Murthy, P. N., and Raboy, V. 2003. Seed phosphorus and inositol phosphate phenotype of barley low phytic acid genotypes. *Phytochemistry* 62:691–706.

Duffus, C. M., and Cochrane, M. P. 1993. Formation of the barley grain: morphology, physiology, and biochemistry. Pages 31–72 in: *Barley: Chemistry and Technology*. A. W. MacGregor and R. S. Bhatty, eds. American Association of Cereal Chemists, St. Paul, MN.

Edney, M. J., Rossnagel, B. G., Endo, Y., Ozawa, S., and Brophy, M. 2002. Pearling quality of Canadian barley varieties and their potential use as rice extenders. *J. Cereal Sci.* 26:295–305.

Ehrenbergerová, J., Belcrediová, N., Pryma, J., Vaculová, K., and Newman, C. W. 2006. Effect of cultivar, year grown, and cropping system on the content of tocopherols and tocotrienols in grains of hulled and hulless barley. *Plant Foods Human Nutr.* 61:145–150.

El-Negoumy, A. M., Newman, C. W., and Moss, B. R. 1977. Chromatographic fractionation and composition of the salt soluble proteins from Hiproly (CI 3947) and Hiproly Normal (CI 4362) barleys. *Cereal Chem.* 54:333–344.

El-Negoumy, A. M., Newman, C. W., and Moss, B. R. 1979. Amino acid composition of total protein and electrophoretic behavior of protein in fractions of barley. *Cereal Chem.* 56:468–473.

Eppendorfer, W. 1968. The effect of nitrogen and sulfur on changes in nitrogenous fractions of barley plants at various stages of growth and on yield and amino acid composition of grain. *Plant Soil* 29:424–438.

Eslick, R. F. 1979. Barley breeding for quality at Montana State University. Pages 2–25 in: *Proc. Joint Barley Utilization Seminar*. Korea Science and Engineering Foundation, Suweon, Korea.

Eslick, R. F. 1981. Mutation and characterization of unusual genes associated with the seed. Pages 864–867 in: *Barley Genetics IV: Proc. 4th International Barley Genetics Symposium*, Edinburgh, UK.

Fastnaught, C. E. 2001. Barley fiber. Pages 519–542 in: *Handbook of Dietary Fiber*. S. S. Cho, and M. L. Dreher, eds. Marcel Dekker, New York.

Fedak, G., and de la Roche, I. A. 1977. Lipid and fatty acid composition of barley kernels. *Can. J. Plant Sci.* 57:257–260.

Fincher, G. B., and Stone, B. A. 1986. Cell walls and their components in cereal grain technology. Pages 207–295 in: *Advances in Cereal Science and Technology*, vol. 8. Y. Pomeranz, ed. American Association Cereal of Chemists, St. Paul, MN.

Franckowiak, J. 1997. Revised linkage maps for morphological markers in barley, *Hordeum vulgare*. *Barley Genet. Newsl.* 26:9–21. Published online at http://www.wheat.pw.usda.gov/ggpages/bgn/.

Gaines, R. L., Bechtel, D. B., and Pomeranz, Y. 1985. A microscopic study on the development of a layer in barley that causes hull–caryopsis adherence. *Cereal Chem.* 62:35–40.

Gardner, K. H., and Blackwell, J. 1974. The structure of native cellulose. *Biopolymers* 13:1975–2001.

Goering, K. J., and Eslick, R. F. 1976. Barley starch: VI. A self-liquefying waxy barley starch. *Cereal Chem.* 53:174–180.

Goering, K. J., Fritts, D. H., and Eslick, R. F. 1973. A study of starch granule size and distribution in 29 barley varieties. *Starch/Stärke* 25:297–302.

Greenberg, D. C. 1977. A diallel cross analysis of gum content in barley (*Hordeum vulgare*). *Theor. Appl. Genet.* 50:141–46.

Greenberg, I. C., and Whitmore, E. T. 1974. A rapid method for estimating the viscosity of barley extracts. *J. Inst. Brew.* 80:31–33.

Groupy, P., Hugues, M., Boivin, J. P., and Amiot, M. J. 1999. Antioxidant compounds and activity of barley (*Hordeum vulgare*) and malt extracts. Pages 445–451 in: *Proc. Cannes EBC Congress*.

Gymer, P. T. 1978. The genetics of the two-row/six-row character. *Barley Genet. Newsl.* 8:44–46. Published on line at http://www.wheat.pw.usda.gov/ggpages/bgn/.

Han, F., Ullrich, S. E., Chriat, S., Menteur, S., Jestin, L., Sarrafi, A., Hayes, P. M., Jones, B. L., Blake, T. K., Wesenberg, D. M., Kleinhofs, A., and Kilian, A. 1995. Mapping of β-glucanase activity loci in barley grain and malt. *Theor. Appl. Genet.* 91:921–927.

Harlan, J. R. 1920. Daily development of kernels of Hanchen barley from flowering to maturity at Aberdeen, Idaho. *J. Agric. Res.* 19:393–429.

Harlan, J. R. 1978. On the origin of barley. Pages 10–36 in: *Barley: Origin, Botany, Culture, Winter Hardiness, Genetics, Utilization, Pests*. Agriculture Handbook 338. U.S. Department of Agriculture, Washington, DC.

Hashimoto, S., Shogren, M. D., Bolte, L. C., and Pomeranz, Y. 1987. Cereal pentosans: their estimation and significance: III. Pentosans in abraded grains and milling by-products. *Cereal Chem.* 64:39–41.

Hejgaard, J., and Boisen, S. 1980. High lysine proteins in Hiproly barley breeding: identification, nutritional significance and new screening methods. *Hereditas* 93:311–320.

Henry, R. J. 1986. Genetic and environmental variation in the pentosan and β-glucan contents of barley, and their relation to malting quality. *J. Cereal Sci.* 4:269–277.

Henry, R. J. 1987. Pentosan and (1–3),(1–4)-β-glucan concentration in endosperm and whole grain of wheat, barley, oats, and rye. *J. Cereal Sci.* 6:253–258.

Henry, R. J. 1988. The carbohydrates of barley grains: a review. *J. Inst. Brew.* 94:71–78.

Hernanz, D., Nunez, V., Sancho, A. I., Faulds, C. A. B., Williamson, G., Bartolome, B., and Gomez-Cordoves, C. 2001. Hydroxycinnamic acids and ferulic acid dehydrodimers in barley and processed barley. *J. Agric. Food Chem.* 49:4884–4888.

Hockett, E. A. 1981. Registration of hulless and short-awned spring barley germ plasm. *Crop Sci.* 21:146–147.

Hockett, E. A., and Nilan, R. A. 1985. Genetics. Pages 187–230 in: *Barley*. Agronomy Monograph 26. D. C. Rasmusson, ed. American Society of Agronomy, Crop Science Society of America, and Soil Science Society of America, Madison, WI.

Hockett, E. A., and Standridge, N. N. 1976. Relationship of agronomic and malt characteristics of isogenic Trail to breeding two- and six-rowed barley. Pages 594–603 in: *Barley Genetics III: Proc. 3rd International Barley Genetics Symposium*, Garching, Germany. H. Gaul, ed. Verlag Karl Thiemig, Munich, Germany.

Hofer, P. J. 1985. The composition and nutritional value of a high-lysine, high-sugar barley (*Hordeum vulgare*). M.S. thesis. Montana State University, Bozeman, MT.

Holtekjølen, A. K., Kinitz, C., and Knutsen, S. H. 2006. Flavanol and bound phenolic acid contents in different barley varieties. *J. Agric. Food Chem.* 54:2253–2260.

Hopkins, R. H., Winer, S., and Rainbow, C. 1948. Pantothenic acid in brewing. *J. Inst. Brew.* 54:264–269.

Ingversen, J., Køie, B., and Doll, H. 1973. Induced seed protein content of barley endosperm. *Experientia* 29:1151–1152.

Ishikawa, N., Ishihara, J., and Masamitsu, I. 1995. Artificial induction and characterization of amylose-free mutants of barley. *Barley Genet. Newsl.* 24:4953.

Jende-Strid, B. 1995. Coordinators report: anthocyanin genes. *Barley Geneti. Newsl.* 24:162–165. Published online at http://www.wheat.pw.usda.gov/ggpages/bgn/.

Jensen, J. 1979. Chromosomal location of one dominant and four recessive high-lysine genes in barley mutants. Pages 89–96 in: *Seed Protein Improvement in Cereals and Grain Legumes*. STI/PUB/496. International Atomic Energy Agency, Food and Agriculture Organization, Vienna, Austria.

Jiang, G., and Vasanthan, T. 2000. Effect of extrusion cooking on the primary structure and water solubility of β-glucans from regular and waxy barley. *Cereal Chem.* 77:396–400.

Karlsson, K. E. 1972. Linkage studies on a gene for high lysine content in Hiproly barley. *Barley Genet. Newsl.* 2:34. Published online at http://www.wheat.pw.usda.gov./ggpages/bgn/.

Kent, N. L., and Evers, A. D. 1994. *Kent's Technology of Cereals*, 4th ed. Elsevier Science, Oxford, UK.

Kerckhoffs, D. A., Brouns, F., Hornstra, G., and Mensink, R. P. 2002. Effect on the human lipoprotein profile of β-glucan, soy protein and isoflavones, plant sterols, garlic and tocotrienols. *J. Nutr.* 132:2494–2505.

Kirkman, M. A., Shewry, P. R., and Miflin, F. J. 1982. The effect of nitrogen nutrition on the lysine content and protein composition of barley seeds. *J. Sci. Food Agric.* 33:115–127.

Kleese, R. A., Rasmussen, D. C., and Smith, L. H. 1968. Genetic and environmental variation in mineral element accumulation in barley, wheat, and soybeans. *Crop Sci.* 8:591–593.

Kleinhofs, A., and Han, F. 2002. Molecular mapping of the barley genome. Pages 31–63 in: *Barley Science: Recent Advances from Molecular Biology to Agronomy of Yield and Quality*. G. A. Slafer, J. L. Molena-Cano, R. Savin, J. L. Araus, and I. Romagosa, eds. Haworth Press, Binghamton, NY.

Köksel, H., Edney, J. J., and Özkaya, B. 1999. Barley bulgur: effect of processing and cooking on chemical composition. *J. Cereal Sci.* 29:185–190.

Lampi, A-M., Moreau, R. A., Piironen, V., and Hicks, K. B. 2004. Pearling barley and rye to produce phytosterol-rich fractions. *Lipids* 39:783–787.

LaBerge, D. E., and Tipples, K. H. 1989. Quality of western Canadian malting barley. Crop Bull. 180. Canadian Grain Commission, Winnipeg, MB, Canada.

Larson, S. R., Young, K. A., Cook, A., Blake, T. K., and Raboy, V. 1998. Linkage mapping of two mutations that reduce phytic acid content of barley grain. *Theor. Appl. Genet.* 97:141–146.

Li, Y. C., Ledoux, D. R., Veum, T. L., Raboy, V., Zyla, K., and Wikiera, A. 2001a. Bioavailability of phosphorus in low phytic acid barley. *J. Appl. Poult. Res.* 10:86–91.

Li, Y. C., Ledoux, D. R., Veum, T. L., Raboy, V., and Zyla, K. 2001b. Low phytic acid barley improves performance, bone mineralization, and phosphorus retention in turkey poults. *J. Appl. Poult. Res.* 10:178–185.

Linde-Laursen, I. 1997. Recommendations for the designation of the barley chromosomes and their arms. *Barley Genet. Newsl.* 26:1–3. Published online at http://www.wheat.pw. usda.gov/ggpages/bgn.

Liu, D. J., and Pomeranz, Y. 1975. Distribution of minerals in barley at the cellular level by x-ray analysis. *Cereal Chem.* 52:620–629.

Liu, D. J., Robbins, G. S., and Pomeranz, Y. 1974. Composition and utilization of milled barley products: IV. Mineral components. *Cereal Chem.* 51:309–316.

MacGregor, A. W., and Fincher, G. F. 1993. Carbohydrates of the barley grain. Pages 73–130 in: *Barley: Chemistry and Technology.* A. W. MacGregor and R. S. Bhatty, eds. American Association of Cereal Chemists, St. Paul, MN.

MacLeod, A. M. 1952. Studies on the free sugars of the barley grain: II. Distribution of the individual sugar fractions. *J. Inst. Brew.* 58:363–371.

MacLeod, A. M. 1953. Studies on the free sugars of the barley grain: IV. Low molecular weight fructosans. *J. Inst. Brew.* 59:462–469.

MacLeod, A. M. 1957. Raffinose metabolism in germinating barley. *New Phytol.* 56: 210–220.

MacLeod, A. M. 1959. Cellulose distribution in barley. *J. Inst. Brew.* 65:188–196.

Marconi, E., Graziano, M., and Cubadda, R. 2000. Composition and utilization of barley pearling by-products for making functional pastas rich in dietary fiber and β-glucans. *Cereal Chem.* 77:133–139.

Mason, J. B., Gibson, N., and Kodicek, E. 1973. The chemical nature of the bound nicotinic acid of wheat bran: studies of nicotinic acid–containing macromolecules. *Br. J. Nutr.* 30:297–300.

Mayler, J. L., and Stanford, E. H. 1942. Color inheritance in barley. *J. Am. Soc. Agric.* 34:427.

Mazza, G., and Gao, L. 2005. Blue and purple grains. Pages 313–350 in: *Specialty Grains for Food and Feed.* E. Abdel-Aal and P. Wood, eds. American Association of Cereal Chemists, St. Paul, MN.

Mazza, G., and Miniati, E. 1993. *Anthocyanins in Fruits, Vegetables, and Grains.* CRC Press, Boca Raton, FL.

McGuire, C. F., and Hockett, E. A. 1981. Effect of awn length and naked caryopsis on malting quality of 'Betzes' barley. *Crop Sci.* 21:18–21.

McLaughlin, P. J., and Weibrauch, J. L. 1979. Vitamin E content of foods. *J. Am. Diet. Assoc.* 75:647–651.

Merritt, N. R. 1967. A new strain of barley with starch of high amylose content. *J. Inst. Brew. (London)* 73:583–585.

Mertz, E. T., Bates, L. S., and Nelson, E. E. 1964. Mutant gene that changes protein composition and increases lysine content of maize endosperm. *Science* 145:279–280.

Miflin, B. J., Field, J. M., and Shewry, P. R. 1983. Cereal storage proteins and their effecrts on technological properties. Pages 255–319 in: *Seed Proteins.* J. Danssant, J. Morse, and J. Vaughan, eds. Academic Press, London.

Miller, S. S., and Fulcher, R. G. 1994. Distribution of (1–3),(1–4)-β-D-glucan in kernels of oats and barley using micro spectrofluorometry. *Cereal Chem.* 71:64–68.

Mohan, D. D., and Axtell, J. D. 1975. Diethyl sulfate induced high lysine mutants in sorghum. In *Proc. 9th Biennial Grain Sorghum Research and Utilization Conference*, Lubbock, TX.

Molina-Cano, J. L., DeTogores, F. R., Royo, C., and Perez, A. 1989. Fast germinating low β-glucan mutants induced in barley with improved malting quality and yield. *Theor. Appl. Genet.* 78:748–754.

Moreau, R. A., Flores, R. A., and Hicks, K. B. 2007. Composition of functional lipids in hulled and hulless barley fractions obtained by scarification and in barley oil. *Cereal Chem.* 84:1–5.

Morell, M. K., Kosar-Hashemic, B., Cmiel, M., Samuel, M. S., Chandler, P., Rahman, S., Buleon, A., Batey, I. L., and Li, Z. 2003. Barley sex6 mutants lack starch synthase IIa activity and contain a starch with novel properties. *Plant J.* 34:173–185.

Morgan, A. G. 1977. The relationship between barley extract viscosity curves and malting quality. *J. Inst. Brew.* 83:231–234.

Morrison, F. B. 1953. *Feeds and Feeding*. The Morrison Publishing Company, Ithaca, New York.

Morrison, W. R. 1988. Lipids. Pages 373–439 in: *Wheat: Chemistry and Technology*, 3rd ed., vol. 1. Y. Pomeranz, ed. American Assoceation of Cereal Chemists, St. Paul, MN.

Morrison, W. R. 1993a. Barley lipids. Pages 199–246 in: *Barley: Chemistry and Technology*. A. W. MacGregor and R. S. Bhatty, eds. American Association of Cereal Chemists, St. Paul, MN.

Morrison, W. R. 1993b. Cereal starch granule development and composition. Pages 175–190 in: *Seed Storage Compounds: Biosynthesis, Interactions and Manipulation*. P. R. Shewry and A. K. Stobart, eds. Oxford University Press, Oxford, UK.

Morrison, W. R., and Scott, D. C. 1986. Measurement of the dimensions of wheat starch granule populations using a Coulter counter with 100-channel analyzer. *J. Cereal Sci.* 4:13–21.

Morrison, W. R., Milligan, T. P., and Azudin, M. N. 1984. A relationship between the amylose and lipid contents of starches from diploid cereals. *J. Cereal Sci.* 2:257–271.

Morrison, W. R., Scott, D. C., and Karkalas, J. 1986. Variation in the composition and physical properties of barley starches. *Starch/Stärke* 38:374–379.

Munck, L. 1972. Improvement of nutrition value in cereals. *Hereditas* 72:1–128.

Munck, L. 1992. The case of high-lysine barley breeding. Pages 573–601 in: *Barley: Genetics, Biochemistry, Molecular Biology and Biotechnology*. P. R. Shewry, ed. C.A.B. International, Wallingford, UK.

Munck, L., Karlson, K. D., Hagberg, A., and Eggum, B. O. 1970. Gene for improved nutritional value in barley seed protein. *Science* 168:985–987.

National Research Council (1984). *Nutrient Requirements of Poultry*, 8th ed. National Academy Press, Washington, DC.

Newman, C. W., and McGuire, C. F. 1985. Nutritional quality of barley. Pages 403–556 in: *Barley*. Agronomy Monograph 26. D. C. Rasmusson, ed. American Society of Agronomy, Crop Science Society of America, and Soil Science Society of America, Madison, WI.

Newman, C. W., and Newman, R. K. 2005. Hulless barley for food and feed. Pages 167–202 in: *Specialty Grains for Food and Feed*. E. Abdel-Aal and P. J. Wood, eds. American Association of Cereal Chemists, St. Paul, MN.

Newman, C. W., Eslick, R. F., and El-Negoumy, A. M. 1982. Comparative feed value of 2-rowed and 6-rowed isotypes of Titan barley. *J. Anim. Sci.* 55(Suppl. 1): 288 (Abstr.).

Newman, C. W., Øverland, M., Newman, R. K., Bang-Olsen, K., and Pedersen, B. 1990. Protein quality of a new high-lysine barley derived from Risø 1508. *Can. J. Anim. Sci.* 70:279–285.

Nilan, R. A. 1964. *The Cytology and Genetics of Barley 1951–1962*. Washington State University, Pullman, WA.

Nilan, R. A. 1974. Barley (*Hordeum vulgare*). Pages 93–110 in: *Handbook of Genetics*. R. C. King, ed. Plenum Press, New York.

Nilan, R. A. 1990. The North American Barley Genome Mapping Project. *Barley Newsl.* 33:112. Published online at http://www.wheat.pw.usda.gov/ggpages/bgn/.

Nilan, R. A., and Ullrich, S. E. 1993. Barley: taxonomy, origin, distribution, production, genetics, and breeding. Pages 1–29 in: *Barley: Chemistry and Technology*. A. W. MacGregor and R. S. Bhatty, eds. American Association Cereal Chemists, St. Paul, MN.

O'Dell, B. L., Boland, A. R-de, and Koirtyohann, A. R. 1972. Distribution of phytate and nutritionally important elements among the morphological components of cereal grains. *J. Agric. Food Chem.* 23:1179–1182.

Osborne, T. B. 1924. *The Vegetable Proteins*, 2nd ed. Longmans, Green, London.

Oser, B. L. 1965. *Hawk's Physiological Chemistry*. McGraw-Hill, New York.

Owen, B. D., Sosulski, F., Wu, K. K., and Farmer, J. M. 1977. Variation in mineral content of Saskatchewan feed grains. *Can. J. Anim. Sci.* 57:679–687.

Parrish, F. W., Perlin, A. S., and Reese, E. T. 1960. Selective enzymolysis of poly-β-D-glucans, and the structure of the polymers. *Can. J. Chem.* 38:2094–2104.

Parsons, J. G., and Price, P. B. 1974. Search for barley (*Hordeum vulgare* L.) with higher lipid content. *Lipids* 9:804–808.

Peterson, D. M. 1994. Barley tocols: effects of milling, malting and mashing. *Cereal Chem.* 71:42–44.

Peterson, D. M., and Qureshi, A. A. 1993. Genotype and environment effects on tocols of barley and oats. *Cereal Chem.* 70:157–162.

Pomeranz, Y. 1982. Grain structure and end-use properties. Pages 107–124 in: *Food Microstructure*, vol. 1. SEM Inc., AMF O'Hare, Chicago.

Pomeranz, Y., Robbins, G. S., Wesenberg, D. M., Hockett, and Gilbertson, J. T. 1973. Amino acid composition of two-rowed and six-rowed barleys. *J. Agric. Food Chem.* 21:218–221.

Powell, W., Caligari, P. D. S., Swanston, J. S., and Jinks, J. L. 1989. Genetic investigations into beta-glucan content in barley. *Theor. Appl. Genet.* 71:461–466.

Price, P. B., and Parsons, J. G. 1974. Lipids of six cultivated barley (*Hordeum vulgare* L.) varieties. *Lipids* 9:560–566.

Price, P. B., and Parsons, J. G. 1975. Lipids of seven cereal grains. *J. Am. Oil Chem. Soc.* 52:490–493.

Price, P. B., and Parson, J. G. 1979. Distribution of lipids in embryonic axis, bran-endosperm and hull fractions of hulless barley and hulless oat grain. *J. Agric. Food Chem.* 27:813–815.

Raboy, V. 1990. Biochemistry and genetics of phytic acid synthesis. Pages 55–76 in: *Inositol Metabolism in Plants*. Wiley-Liss, New York.

Raboy, V., and Cook, A. 1999. An update on ARS barley low phytic acid research. *Barley Genet. Newsl.* 29:33–35. Published online at http://www.wheat.pw.usda.gov/ggpages/bgn/.

Raboy, V., Young, K. A., Dorsch, J. A., and Cook, A. 2001. Genetics and breeding of seed phosphorus and phytic acid. *J. Plant Physiol.* 158:489–497.

Ragaee, S., Abdel-Aal, E. M., and Noaman, M. 2006. Antioxidant activity and nutrient composition of selected cereals for food use. *Food Chem.* 98:32–38.

Ramage, R. T. 1985. Cytogentics. Pages 127–154 in: *Barley*. Agronomy Monograph 26. D. C. Rasmusson, ed. American Society of Agronomy, Crop Science Society of America, and Soil Science Society of America, Madison, WI.

Shewry, P. R. 1993. Barley seed proteins. Pages 131–197 in *Barley: Chemistry and Technology*. A. W. MacGregor and R. S. Bhatty, eds. American Association of Cereal Chemists, St. Paul, MN.

Shewry, P. R., and Darlington, H. 2002. The proteins of the mature barley grain and their role in determining malting performance. Pages 503–521 in: *Barley Science: Recent Advances from Molecular Biology to Agronomy of Yield and Quality*. G. A. Slafer, J. L. Molina-Cano, R. Savin, J. L. Araus, and I. Romagosa, eds. Haworth Press, New York.

Shewry, P. R., and Tatham, A. S. 1987. Recent advances in our understanding of cereal seed protein structure and functionality: comments. *Agric. Food Chem.* 1:71–93.

Shewry, P. R., and Tatham, A. S. 1990. The prolamin storage proteins of cereal seeds: Structure and evaluation. *Biochem. J.* 267:1–12.

Shewry, P. R., Bright, S. W. J., Burgess, S. R., and Miflin, B. J. 1984. Approaches to improving the nutritional value of barley seed proteins. Pages 227–240 in: *Use of Nuclear Techniques for Cereal Grain Improvement*. STI/PUB/664. International Atomic Energy Agency, Vienna.

Smith, L. 1951. Cytology and genetics of barley. *Bot. Rev.* 17:1–355.

Sogaard, B., and Wettstein-Knowles, P. von. 1987. Barley genes and chromosomes. *Carlsberg Res. Commun.* 52:123–196.

Stahl, A., Persson, G., Johannsson, L.-A., and Johannsson, H. 1996. Breeding barley for functional food starch. Pages 95–97 in: *Proc. 5th International Barley Genetics Symposium*, University Extension Press, University of Saskatchewan, saskatoon, SK, Canada.

Sugiura, S. H., Raboy, V., Young, K. A., Dong, F. M., and Hardy, R. W. 1999. Availability of phosphorus and trace elements in low-phytate varieties of barley and corn for rainbow trout (*Oncorhynchus mykiss*). *Aquaculture* 170:285–296.

Swanston, J. S. 1997. Waxy starch barley genotypes with reduced β-glucan contents. *Cereal Chem.* 74:452–455.

Swanston, J. S., Ellis, R. P., and Tiller, S. A. 1997. Effects of the waxy and high-amylose genes on the total beta-glucan and extractable starch. *Barley Genet. Newsl.* 27:72–74. Published online at http://www.wheat.pw.usda.gov/ggpages/bgn/.

Takahashi, R. 1955. The origin and evolution of cultivated barley. *Adv. Genet.* 7:227–266.

Takeda, Y., Shirasaka, K., and Hizukuri, S. 1984. Examination of the purity and structure of amylose by gelpermentation chromatography. *Carbohydr. Res.* 132:83–92.

Tester, R. F., and Morrison, W. R. 1992. Swelling and gelatinization of cereal starches: III. Some properties of waxy and normal nonwaxy starches. *Cereal Chem.* 69:654–658.

Truscott, D. R., Newman, C. W., and Roth, N. J. 1988. Effects of hull and awn type on the feed value of Betzes barley. I. Chemical and physical characteristics of the kernels. *Nutr. Reps Int.* 37:681–693.

Ullrich, S. E. 2002. Genetics and breeding of barley feed quality. Pages 115–142 in: *Barley Science: Recent Advances from Molecular Biology to Agronomy of Yield and Quality.* G. A. Slafer, J. L. Molina-Cano, R. Savin, J. L. Araus, and I. Romagosa, eds. Haworth Press, Binghamton, NY.

Ullrich, S. E., Clancy, J. A., Eslick, R. F., and Lance, R. C. M. 1986. Beta-glucan content and viscosity of waxy barley. *J. Cereal Sci.* 4:279–285.

USDA (U.S. Department of Agriculture, Agricultural Research Service) 2005. USDA nutrient data base for standard reference, release 18. Nutrient data laboratory home page, Published online at http://www.nal.usda.gov/fnic/foodcomp.

Vaculová, K., Ehrenbergerová, J., Nemejc, R., and Pryma, J. 2001. The variability and correlations between the content of vitamin E and its isomers in hybrids of the F2 generation of spring barley. *Acta Univ. Agric. et Silva. Mendel. Brun.* XLIX 1, 1–9.

Veum, T. L., Ledoux, D. R., Bollinger, D. W., Raboy, V., and Cook, A. 2002. Low-phytic acid barley improves calcium and phosphorus utilization and growth performance in growing pigs. *J. Anim. Sci.* 80:1–8.

Viëtor, R. J., Angelino, S. A. G. F., and Voragen, A. G. J. 1993. Structural features of arabinoxylans from barley and malt cell wall material. *J. Cereal Sci.* 15:213–222.

Voragen, A. G. J., Viëtor, R. J. and Angelino, S. A. G. F. 1992. Composition and properties of arabinoxylans in barley and malt. Pages 40–46 in: *Barley for Food and Malt. ICC/SCF International Symposium.* Swedish University of Agricultural Science, Uppsala, Sweden.

Walker, J. J., and Merritt, N. R. 1969. Genetic control of abnormal starch granules and high amylose content in a mutant of Glacier barley. *Nature (London)* 221:482–483.

Wang, L. 1992. Influences of oil and soluble fiber of barley grain on plasma cholesterol concentrations in chicks and hamsters. Ph.D. dissertation. Montana State University, Bozeman, MT.

Wang, L., Xue, Q., Newman, R. K., Newman, C. W., and Jackson, L. L. 1993. Tocotrienol and fatty acid composition of barley oil and their effects on lipid metabolism. *Plant Foods Hum. Nutr.* 43:9–17.

Weaver, C. M., Chen, P. H., and Rynearson, S. L. 1981. Effects of milling on trace element and protein content of oats and barley. *Cereal Chem.* 58:120–124.

Western Regional Project-166. 1984. Variation in the compostion of Western grown barley. *Proc. West. Sect. Amer. Soc. Anim. Sci.* 35:163–165.

Wettstein-Knowles, P. von. 1992. Cloned and mapped genes: current status. Pages 73–98 in: *Barley: Genetics Biochemistry, Molecular Biology and Biotechnology.* P. R. Shewry, ed. C.A.B. International, Wallingford, UK.

Wood, P. J., Weisz, J., Beer, M. V., Newman, C. W., and Newman, R. K. 2003. Structure of $(1-3)(1-4)$-β-D-glucan in waxy and nonwaxy barley. *Cereal Chem.* 80:329–332.

Woodward, J. R., and Fincher, G. B. 1983. Water soluble barley β-glucans: fine structure, solution behavior, and organization in the cell wall. *Brew. Dig.* 58:28–32.

Woodward, J. R., Fincher, G. B., and Stone, B. A. 1983a. Water-soluble (1–3)(1–4)-β-D-glucans from barley (*Hordeum vulgare*) endosperm: I. Physicochemical properties. *Carbohydr. Polym.* 3:143–156.

Woodward, J. R., Fincher, G. B., and Stone, B. A. 1983b. Water-soluble (1–3)(1–4)-β-D-glucans from barley (*Hordeum vulgare*) endosperm: II. Fine structure. *Carbohydr. Polym.* 3:207–225.

Woodward, R. W., and Thieret, J. W. 1953. A genetic study of complementary genes for purple lemma, palea, and pericarp in barley. *(Hordeum vulgare L.). Agron. J.* 45:182–185.

Xue, Q. 1992. Influence of waxy and high-amylose starch genes on the composition of barley and the cholesterolemic and glycemic responses in chicks and rats. Ph.D. dissertation. Montana State University, Bozeman, MT.

Xue, Q., Wang, L., Newman, R. K., Newman, C. W., and Graham, H. 1997. Influence of the hulless, waxy starch, and short-awn genes on the composition of barleys. *J. Cereal Sci.* 26:251–257.

Yeung, J., and Vasanthan, T. 2001. Pearling of hull-less barley: product composition and gel color of pearled barley flours as affected by degree of pearling. *J. Agric. Food Chem.* 49:331–335.

Zale, J. M., Clancy, J. A., Jones, B. L., Hayes, P. M., and Ullrich, S. E. 2000. Summary of barley malting quality QTLs mapped in various populations. *Barley Genet. Newsl.* 30:44–54. Published online at http://www.wheat.pw.usda.gov/ggpages/bgn/.

Zheng, G. H., Rossnagel, B. G., Tyler, R. T., and Bhatty, R. S. 2000. Distribution of β-glucan in the grain of hulless barley. *Cereal Chem.* 77:140–144.

Zupfer, J. M., Churchill, K. E., Rasmusson, D. C., and Fulcher, R. 1998. Variation in ferulic acid concentration among diverse barley cultivars measured by HPLC and microspectrophotometry. *J. Agric. Food Chem.* 46:1350–1354.

5 Barley Processing: Methods and Product Composition

INTRODUCTION

As with other cereal grains, barley requires one or more processing procedures in order to reduce or change the kernels into a usable, edible form. Modern processing methods do one or more of the following: alter physical form or particle size, isolate and concentrate specific parts of the grain, improve mouthfeel and palatability, alter nutrient digestibility, and improve product stability or shelf life. Some form of milling, either dry or wet, is the processing most commonly applied to cereal grains. With the exception of malting and extrusion, we deal primarily with dry milling processes in this chapter. Dry milling encompasses a variety of abrasion, grinding, flaking, and separation procedures utilizing numerous types of specialized mills. The earliest processing method used to prepare cereal grains such as barley was a crude form of grinding or crushing the kernels with hand-held stones in the style of a mortar and pestle. As civilization developed, a more advanced form of stone milling was developed using a quern, which consisted of two stones, a rotating "capstone" and a stationary "netherstone." Rotary stone mills, whose invention is attributed to the Romans in the second century B.C., were used almost exclusively to grind grain until the mid nineteenth century. The first stone mills were hand operated, then powered by animals, and were later water powered (Kent and Evers 1994; Owens 2001).

A multitude of food products can be produced from barley using one or more of several processing or milling procedures that are available in commerce. Barley may be milled by blocking (dehulling), pearling, flaking, cutting, grinding, roller milling, and further processed by air classifying, sieving, extruding, and infrared heating. Malting is a separate process used to prepare barley as a specialty food additive and in alcoholic beverage production. These processes will be discussed briefly along with products produced and nutrient composition of these products. The purpose of this chapter is not to give detailed analyses of the processes, but

Barley for Food and Health: Science, Technology, and Products,
By Rosemary K. Newman and C. Walter Newman
Copyright © 2008 John Wiley & Sons, Inc.

to provide the reader with a basic understanding of how barley is transformed from a raw grain into edible food products with enhanced nutrient profiles and to provide credible references to obtain detail information.

WHOLE-GRAIN PROCESSING

Cleaning, Sizing, and Conditioning

The first step in processing barley is cleaning to remove foreign grains, weed seed, and impurities such as dirt, sand, dust, straw, chaff, and string, and to remove broken kernels. This is accomplished by use of specialized equipment, including screens, magnets, de-stoners, gravity tables, separators, and scourers, and aspiration with air currents. A process of sieving with selective screens is effective in separating small, less hazardous impurities once larger contaminants are removed. Specific-gravity separations utilizing air currents may also be accomplished once the partially decontaminated grain leaves the sieving process. This will separate lighter material, such as dust, chaff, straw, and string, from the grain (Wingfield 1980; Kent and Evers 1994).

Kernels preferred for milling are more or less uniform in size, white or creamy in color, of medium hardness, and thin-hulled (if hulled barley). Depending on the milling process to be used, barley kernels may be sorted for desired size and shape with rejected kernels channeled into other, more appropriate processes or utilized for animal feed. Sizing and separation may be achieved by disk separators such as the Carter–Day (Carter-Day Company, Minneapolis, Minnesota). Selection of the desired size of kernel is based on length, as the width of the kernel is of no consequence. Disks mounted on a horizontal axle are positioned in a trough that receives the grain mixture to be sorted. Each disk has a series of indentations arranged concentrically, conforming to the size of kernel desired. Acting as pockets, the indentations in the disk pick up kernels of the same size and lift them as they are rotated. The kernels fall out of the disk at the apex of the rotation into a collection channel, where they are moved to a point for accumulation and further processing (Kent and Evers 1994).

In most processing centers, cereal grains are no longer cleaned by washing with water, although there is a pretreatment process called *damping*, which means the controlled addition of moisture. This process is often referred to as *conditioning* or *tempering* of the grain, which alters some of the physical properties. Conditioning is not a single process for all grains, being specific for the individual grain and the milling process. The volume and temperature of the water added will have major effects on processing. Considerably less is known about conditioning barley for milling compared to wheat and oats. Recent research has focused on understanding conditioning of barley, especially in the production of flour through roller milling; however, recommendations are limited for barley milling other than for pearling. According to Kent and Evers (1994), the moisture content of barley grain should be adjusted to 15% through tempering, and allowed to rest or lie for 24 hours prior to pearling. For producing barley flour through roller milling, these authors recommended conditioning moisture levels for pearl barley and unpearled hulless grain of 13 and 14%, respectively, followed by 48 hours of rest. Similar conclusions were arrived at by Andersson et al. (2003) when roller-milling hulless barleys with a Bühler laboratory mill. These authors compared no conditioning with 11.3, 14.3, 16.3, and 17.8% moisture, followed by mixing and allowing to rest overnight. The best milling performance was obtained with barley conditioned to 14.3% moisture. Bhatty (1999a) roller-milled nonwaxy hulless barleys into two fractions, bran and flour, at moisture conditioning levels ranging from 5 to 16%. He found the greatest total yield of flour at the lowest level of moisture, but the highest levels of starch in the flour were produced at the highest level of moisture. It is most likely that preferred conditioning requirements for roller milling barley will depend on the desired end product as well as the genotype and protein content of the barley. Malting or modification of barley involves considerably greater amounts of added moisture during steeping and germination than is required for conditioning. A minimum of 30% and up to 50% water is added to facilitate germination in the malting process.

Blocking and Pearling

Blocking and pearling are processes whereby the outer barley grain tissues are gradually removed by an abrasive scouring action to produce an edible product for food. *Blocking*, sometimes referred to as *hulling (dehulling)*, is usually the next processing step after cleaning. Blocking is designed to remove the tough, fibrous, and largely indigestible hull that adheres strongly to the kernel. The hull represents 10 to 13% of the dry weight of the kernel, but under some conditions, blocking in a commercial dehulling or pearling machine may remove as much as 20% of the kernel weight. Dehulling machines are comprised of circular abrasive stones rotating on a vertical or horizontal axis in a cylindrical chamber surrounded by screens. The cylinder is strongly aspirated, and the degree of abrasion is governed by the counterpressure applied at the outlet, the abrasiveness of the stones, and the distance between the rotor and the surrounding screens (Vorwerck 1992). The *pearling* process that follows blocking is commonly done in three or more stages. The latter stage or stages are sometimes referred to as *polishing*. The degree of pearling, typically stated as a percentage, refers to the amount of the kernel removed during processing. The term *degree of pearling* is defined differently in different geographic regions. In the United States and Canada, for example, 30% pearled means that 30% of the original weight is removed, leaving 70% as pearled barley. In Japan, the same product is termed 70% pearled (Yeung and Vasanthan 2001). The material removed in the pearling process is referred to as *pearling flour*, *fines*, or *pearlings*. Processing of any type will in most cases alter the composition and ratio of nutrients in the final product(s) given the varying concentrations of nutrients that occur in the separate tissues of the kernel. Pearling time to achieve a specific product concentration will also vary with the cultivar, especially between hulled and hulless cultivars and waxy and nonwaxy barley. Differences in pearling characteristics within the same cultivar may be due to environmental factors affecting kernel size and hardness. Pearling hulless barley is optional, depending on the proposed end use and the targeted market. If hulless barley is to be ground into a whole meal or milled into fractions, pearling may not be necessary except for certain targeted markets. Pearling of hulled and hulless barley is desirable where color and uniformity of kernel size are important.

Barley quality requirements for the rice extender market in Southeast Asia were described by Edney et al. (2002). Those consumers demand symmetrical grains that resemble cooked rice in size and color. To achieve a desirable shape and size, the barley is pearled, split, and pearled (polished) again. These processes require barley grains with special texture, size, and color to meet specifications. These researchers evaluated a total of 30 barley varieties of different genotypes for rice extender processing and developed a ranking system for overall desirability. They measured pearling time, whiteness, brightness, crease width, percent of broken kernels, percent of particle size, kernel size, percent of protein, rigidity value, and percent of steely kernels. Distinct differences affecting the desirability for pearling were noted between genotypes, particularly in regard to steely kernels, broken kernels, and rigidity value. Waxy genotypes had better overall characteristics than those of hulled or hulless normal starch cultivars. Hulless genotypes had better color than hulled genotypes but poorer endosperm texture. Brightness appeared to be a good parameter for screening barley samples for good rice extender quality.

One of the earliest reports on nutrient composition of pearl barley and pearlings was that of Norman et al. (1965). This study revealed increased protein levels in the pearlings, with a concomitant lowering of protein in the pearl barley compared to the whole grain. Three malting barley cultivars were pearled to remove above 50% of the kernel. Average protein of the original kernels was 11.0% and that of the pearled kernels was 7.2%, a decrease of 34.5%. Protein concentration in the pearling flour was least in the first fraction, which consisted mostly of the hulls. Protein concentration was greatest in the second and third fractions and decreased in the fourth fraction, with a final decrease in the fifth fraction down to near the level in the intact kernel. In a study principally to measure trace mineral levels in pearled barley, Weaver et al. (1981) reported a decrease in protein in the pearled kernel by only 8% (13.2 to 12.2%), clearly demonstrating the difference in results of pearling on nutrient concentration that may occur between laboratories or commercial plants using different equipment and possibly, barley cultivars. Levels of iron, zinc, manganese, copper, chromium, and nickel in whole kernels and pearled kernels were 45.2, 29.2; 34.4, 23.4; 27.2, 19.0; 6.3, 5.4; 2.1, 2.0; and 0.9, 0.6 mg/kg, respectively. Iron, zinc, and manganese levels were reduced in the pearlings by about 33% compared to the intact kernels, while levels of copper, chromium, and nickel were not greatly changed by pearling.

Bach Knudsen and Eggum (1984) researched the composition nutritive value of botanically defined milled fractions (pearlings) of hulled barley by successive decortication (pearling) of the kernel. Soluble dietary fiber and β-glucan levels increased in the pearlings up to about 30% pearling and then remained stable, while the starch level showed a continued increase with pearling up to 80%. In a pearling study reported by Pedersen et al. (1989), the hull and seed coat were in the first pearling fraction (11% of the total), the germ and aleurone made up the second pearling fraction (11 to 25% of the total), and the remaining 75% found in the third pearling fraction was composed primarily of the endosperm. Nutrient composition of the pearling fractions represented the component parts of the kernel from which the fractions originated. Sumner et al. (1985) compared the pearling characteristics of waxy and nonwaxy hulled and hulless barley at three extraction rates—high, medium, and low—resulting in yields of 84, 71, and 56% pearl barley, respectively. Protein percentages were decreased in the pearl barley and increased in the pearlings at each successive extraction level, with opposite effects on the percentage of starch. Percentages of ash and fat were also decreased in the pearled barley with concomitant increases in the pearlings, and were highest in pearlings obtained from 71% extraction. This was indicative of the mineral and oil contents of the outer layers of the kernel and germ, respectively.

Bhatty and Rossnagel (1998) compared the composition of 12 Canadian and seven Japanese barleys pearled to 55% extraction in a Satake mill. The Canadian barleys were hulless, whereas only two of the Japanese barleys were hulless. Both groups of cultivars responded to pearling in a similar manner, although differences were found in the ratio of protein and starch between the Canadian and Japanese barleys. Fiber type and level differed due to the presence of hull in some of the Japanese barleys. Protein, ash, and total dietary fiber (TDF) decreased in the pearled barley kernel compared to the whole-grain kernels, whereas starch level and extract viscosity increased. Soluble dietary fiber (SDF) and β-glucan levels were essentially the same in whole and pearl barley. The data of Bhatty and Rossnagel (1998) shown in Table 5.1 are representative of those reported in most pearling studies (Norman et al. 1965; Bach Knudsen and Eggum 1984; Sumner et al. 1985; Pedersen et al. 1989). In this study, three of the 12 Canadian barleys were waxy, which had the highest levels of SDF and β-glucans in the pearled kernels of all barleys, including the Japanese cultivars.

Zheng et al. (2000) and Yeung and Vasanthan (2001) compared pearling characteristics of nonwaxy and waxy hulless barleys. In both of these studies, β-glucan levels in the pearled kernels of nonwaxy barley showed a sharp increase up to about 25% pearling, followed by a gradual decrease toward the center of the kernel. These data suggested highest concentrations of β-glucan in aleurone, sub-aleurone, and outer starchy endosperm of nonwaxy barley. In contrast, a gradual increase in β-glucans in the pearled kernels of waxy barley indicated a positive concentration gradient of these compounds toward the center of the kernel. These data and those of others confirm earlier reports, such as those of Ullrich et al. (1986) and Xue et al. (1997), concerning the positive relationship between the waxy gene and β-glucan levels in barley.

Pearled barley and pearling by-products were obtained from two commercial stocks of hulled nonwaxy barley of English and Italian origin (Marconi et al. 2000). The barleys were pearled using an industrial process in five consecutive

TABLE 5.1 Composition of Japanese and Canadian Whole Barley Kernels (WK) and Pearled Barley Kernels (PK) and Extract Viscosity, Dry Matter Basis

	Japanese Barley ($n = 7$)		Canadian Barley ($n = 12$)	
Item	Mean	Range	Mean	Range
Protein (g/100 g)				
WK	10.5	8.7–12.6	15.5	12.8–17.8
PK	6.6	5.8–8.1	9.0	8.1–10.2
Ash (g/100 g)				
WK	2.1	1.7–2.8	2.0	1.8–2.2
PK	0.6	0.5–0.6	0.6	0.5–0.6
Starch (g/100 g)				
WK	71.5	67.4–77.3	67.5	60.1–74.5
PK	88.7	83.6–91.2	81.4	69.1–87.6
TDF[a] (g/100 g)				
WK	16.9	13.6–21.0	12.9	11.0–16.6
PK	8.8	7.5–9.8	7.3	6.5–9.6
SDF[b] (g/100 g)				
WK	4.1	3.6–5.2	3.6	3.1–6.4
PK	3.9	2.5–5.0	3.4	2.3–6.8
βG[c] (g/100 g)				
WK	4.6	3.8–5.4	5.0	4.1–8.0
PK	4.7	3.5–5.6	4.9	3.9–7.7
Viscosity (cP)				
WK	9.0	2.9–32.9	20.0	7.1–469.0
PK	14.0	3.4–49.8	35.9	18.9–786.9

Source: Bhatty and Rossnagel (1998).
[a]TDF, total dietary fiber.
[b]SDF, soluble dietary fiber.
[c]βG, β-glucan.

steps, removing a total of 37 and 46% of the kernel weight of the English and Italian barleys, respectively. The first two pearling fractions, representing about 10% of the kernel weight, were very high in TDF and insoluble dietary fiber (IDF) and low in β-glucans, components that are representative of fiber and β-glucan levels in the hull. Protein concentration in the first fraction of both barleys was the lowest of the five, but increased in fractions 2 to 4 with a slight decrease in the final fraction. Pearling did not always affect the β-glucan content of pearl barley significantly, although a general trend was found of increased β-glucan levels with each pearling in both the English and Italian barley. These data are shown in Table 5.2. These results are somewhat in contrast to the data reported by Zheng et al. (2000) for nonwaxy barley, although the latter research was with hulless barley, which may react differently from hulled barley. Protein, fat, ash, fiber, and digestible carbohydrate (starch and sugars) levels in pearlings, and pearled barley obtained with the industrial pearling process, followed the pattern reported

TABLE 5.2   Composition of English and Italian Whole Barley Kernels (WK), Pearled Barley Kernels (PK), and Five Successive Pearling By-products (BP-1-5) (g/100 g), Dry Matter Basis

Item	Protein	Lipid	Ash	DC[a]	TDF[a]	IDF[a]	SDF[a]	βG[a]
English barley								
WK	12.4	3.5	2.3	59.7	22.1	16.7	5.4	4.3
PK[b]	9.5	1.8	1.0	74.8	12.9	7.5	5.4	4.5
BP-1	8.2	2.9	8.1	7.4	73.4	72.3	1.1	1.0
BP-2	15.1	7.3	6.5	8.7	62.4	59.3	3.1	2.0
BP-3	18.3	8.4	6.1	22.8	44.4	40.0	4.4	3.3
BP-4	19.4	7.1	4.0	38.1	31.4	25.9	5.5	4.5
BP-5	16.0	4.6	2.4	53.2	23.8	17.2	6.6	5.1
Italian barley								
WK	11.7	3.5	2.6	61.8	20.4	15.4	5.0	3.7
PK[c]	8.3	1.5	1.1	80.0	9.2	4.7	4.5	4.2
BP-1	12.1	6.9	8.4	8.0	64.6	62.7	1.9	1.1
BP-2	17.8	7.2	6.6	16.7	51.7	48.8	2.8	2.0
BP-3	19.8	8.6	5.1	33.4	33.1	28.6	4.6	3.4
BP-4	18.4	7.0	4.1	44.4	26.1	20.3	5.8	3.6
BP-5	14.0	4.3	2.8	57.3	21.6	15.5	6.1	4.4

Source: Marconi et al. (2000).

[a]DC, digestible carbohydrates (starch and sugar); TDF, total dietary fiber; IDF, insoluble dietary fiber; SDF, soluble dietary fiber; βG, β-glucan.
[b]62.9% pearled.
[c]54.1% pearled.

earlier in products produced in laboratory mills. These researchers concluded that pearling rates of 30 to 40% were the most desirable to produce equilibrium between nutritional and physical properties of pearled barley.

The primary food product of the pearling process is pearl barley. The by-product, pearlings or fines, have not been used in human foods to any great extent on a commercial basis. However, the possibility exists of including pearlings in various food products, such as muffins and cookies, to increase the levels of protein, fiber, and lipid components. The pearling process can be used to concentrate metabolically important compounds and other essential nutrients, such as barley oil, which contains significant amounts of tocopherol and tocotrienol. In a study reported by Wang et al. (1993) pearling flour from waxy barley representing 20% of the original kernel weight contained high concentrations of oil (81.5 g/kg), tocopherol (35.4 mg/kg), and tocotrienols (115.8 mg/kg). The tocotrienols were concentrated 2.7 times that found in the whole grain. Moreau et al. (2007) found that the average oil content of pearlings representing 12 to 15% of the kernel of two hulled and two hulless barleys was about 3.4 times that found in the whole grain. These researchers found similar amounts of oil (62.2 g/kg) in the pearlings of these barleys as reported by Wang et al. (1993), although they reported much higher levels of tocopherol (132.6 mg/kg) and tocotrienol (224.8 mg/kg). The different findings between the two studies may in part be explained by differences in analytical techniques, barley cultivars, and method of sample preparation. In the study reported by Wang et al. (1993), the pearlings were produced with a model 6K572 laboratory-type barley pearler (Dayton Electric Manufacturing, Chicago, Illinois), whereas Moreau (2007) used a laboratory seed scarifier (Forsberg, Thief River Falls, Minnesota). Phytosterols as well as the vitamin E components have been shown to have positive health benefits. Lampi et al. (2004) reported that pearling fines of hulless barley contained about 2 mg/g phytosterol compounds, which was at least double that in the whole grain. In addition to the evaluation of barley pearlings for tocols, Moreau et al. (2007) examined the phytosterol contents. They reported an average of 2.35 mg/g total phytosterols in the pearlings, a 209% increase over the level in the whole barley grain.

Commercially pearled barley usually represents approximately 60 to 70% of the original barley kernel and is the typical product used for soups and pilafs. It is also the material for further processing, including grinding, flaking, cutting, and roller milling. Blocked and pearled barley may be used for soups, stews, pilafs, and other dishes and also as a rice extender, as noted without further processing. Pearled barley may be lightly cooked or roasted through various processes to produce "quick-cooking" barley products. Pearling fractions can be produced utilizing various extraction rates that vary widely in their physical, chemical, and functional characteristics. Thus, the various layers removed as by-products in the pearling process can be utilized in formulating numerous end products, especially for the production of functional foods with defined health benefits. When used as a rice extender, pearled barley adds β-glucans and increases the SDF significantly in recipes with rice.

Most experimental studies of pearling barley generally used laboratory-scale pearling machines such as that from Seedburo Equipment Co., Chicago, Illinois or Satake, Tokyo, Japan. The laboratory equipment used in research produces pearl barley and pearlings that are comparable to those produced by large commercial pearling equipment, as evidenced by the similar findings reported by Marconi et al. (2000), thereby negating the need for large quantities of barley for initial evaluations.

Grinding

Grinding of cereal grains in the broad sense can range from simple abrasion, producing whole-meal flour in a stone mill, to a complex mechanical system utilizing various methods for separation and regrinding. Grinding is a form of dry milling for the production of whole-grain flour or meal that may be used as received from the grinder or separated into fractions having different particle size and/or density using air classification and/or sieving techniques.

Stone mills developed by the Romans in the second century B.C. were used for over 1000 years in Europe to grind barley into whole meal and flour. A typical stone mill consists of at least two abrasive stone disks positioned on a vertical axis through a hole in the center of the stones. The stones may be as large as 1.2 m in diameter and as thick as 0.3 m. The stone disks may have several grooves extending from the center hole to the outer edges. The grooves facilitate movement of the ground grain toward the edge of the stone. Grain is fed through the center hole as the upper stone is rotated in close contact with lower stone. As the grain is mashed or ground it moves outward through the grooves in the lower stone to the periphery, where it is collected. Advanced stone mills generally have three sets of stone disks. The first cracks the hull, which is removed by a fan mechanism. A second set of disks produces coarse-meal flour, and a third set grinds the meal to fine flour, which may be sieved to remove large particles. Mostly stone mills were replaced by roller mills in the mid-nineteenth century, except for those mills that produce stone milled barley, wheat, and other grains for specialty markets (Kent and Evers 1994; Owens 2001). The Barony Mill at Birsay (www.birsay.org.uk/baronymill.htm) near Kirkwall in the Orkney Islands of Scotland is still in operation. It is a water-powered stone mill that specializes in producing whole-meal barley flour from Bere barley, an ancient cultivar grown in Scotland for at least 2000 years (Theobald et al. 2006). Stone milling is an effective method of producing whole-grain products when complete separation milling or concentration of endosperm for flour is not required for the final product.

In modern processing, hammer mills and pin mills are used more commonly to grind whole-grain or pearled barley for direct use or prior to further processing. A *hammer mill* is a machine designed to shred material, including grain, into fine particles. Hammer mills include a delivery system to introduce grain into the path of rotating hammers. These mills consist of a steel drum or housing containing a rotor mounted on a horizontal shaft on which pivoting free-swinging hammers are suspended from rods running parallel to the shaft and through the rotor disks. The rotor is spun at high speed inside the drum while grain is metered into the grinding area. The grain is contacted there by the hammers, reducing the particle size and forcing the particles through a perforated screen into a discharge opening. Optimal design and placement of the hammers provide maximum contact with the grain. The ground grain is removed either by gravity or is air-assisted using a blower system through a discharge vent. Small screen openings and high-speed rotation of the hammers increase the effectiveness of particle size reduction. The design and placement of hammers are determined by operating parameters such as rotor speed, motor horsepower, and open area in the screen. Rotor speed and the amount of open area in the screen are the primary determinants of grinding efficiency and finished particle size. The screen must be designed to maintain its integrity and provide the greatest amount of open area. Screen openings (holes) that are aligned in a $60°$ staggered pattern optimize open area while maintaining screen strength. Hammer mills are capable of producing a wide range of particle sizes, which will generally be spherical, with a surface that appears polished, although particle size will be less uniform than with some other methods of grinding (Wikipedia 2007).

Pin milling is a means of grinding, sizing, deagglomerating, and homogenizing various materials, including cereal grains. Pin milling is accomplished with a pinned disk grinder, consisting of two steel disks mounted on a vertical or horizontal axis inside a steel chamber. Each disk is studded on the inward-facing surface of the disk with projecting steel pins arranged in intermeshing concentric rings. One disk is stationary (the "stator" disk), while the other rotates, putting grain particles in contact with the pins. A metered amount of grain is fed into the grinding chamber at a steady rate between the two disks through a centrally located inlet. The grain is propelled centrifugally by air currents into the impact zone, where it is subjected to multiple contacts with the pins, resulting in particle size reduction. A pin mill works purely on impact, with the rotor speed controlling the final particle size. The faster the rotor speed, the greater the impacts and the smaller the output particle size will be, and conversely, the slower the rotor speed, the coarser the final product. Typical grinding results are a fairly even bell-shaped curve of particle size that can be moved toward fine or coarse particle sizes with incremental adjustments of the rotor speed. A pin mill is air-swept, which means that a positive airflow is generated from the inlet to the discharge, requiring separation of the air and product at this point (Kent and Evers 1994; Evans 2007). Pin mills are sometimes referred to as *centrifugal impact mills*. In essence, pin milling and hammer milling techniques are based on the same principle: effecting particle size reduction by impact with rotating steel bars, hammers, or pins. Pin mills do not utilize a screen as is the case with hammer mills, and pin mills can produce a finer, more uniform grind.

The hammer milling of hulled barley, commonly used to prepare feed for livestock, is not recommended for foods, as the hull is shredded and difficult to remove completely, even with sieving and aeration. Whole-grain barley flour may be made by hammer milling cleaned hulless barley or from blocked and pearl barley. Small samples may be prepared (ground) under laboratory conditions for analytical purposes utilizing laboratory-scale grinding equipment that mimics commercial grinders.

Roller Milling

The advent of roller milling spelled the end of large-scale stone milling of barley and other grains. Traditionally, *roller milling* was developed to fractionate wheat into "white" flour for baking purposes. *White flour* is defined as "starchy endosperm in the form of particles small enough to pass through a 140 μm aperture flour sieve" (Kent and Evers 1994). Unless otherwise specified, further reference to wheat flour or flour refers to white flour. In the commercial flour milling industry, the criteria for wheat flour will vary to some degree depending on the product specifications desired by a particular mill or required for a particular market. The basic objectives of roller milling are threefold: to separate the endosperm from the bran and embryo, to reduce the maximum amount of endosperm to flour fineness, and to obtain maximum extraction of flour from the grain. The number of parts of flour by weight produced per 100 parts of grain milled is called the *flour yield* or *percentage extraction rate*. This ranges from 72 to 80% for wheat, worldwide. In roller milling, kernels and broken grain particles are subjected to shear and compression when passing between a series of fluted (corrugated) and smooth rotating rollers followed by sieves arranged in succession. The two steps in roller milling are the break system and the reduction system, each of which is followed by sieving to remove the finer particles, producing a composite flour and milling by-products. The *break* step splits the kernels open, exposing the starchy endosperm while crushing the outer layers. The *reduction* system consists of a series of rollers and sifters in sequence used to reduce the particle size of the various kernel parts. These steps are repeated until the flour quality desired is obtained. By-products are those fractions that can yield no appreciable amount of useful flour by repeated rolling and sieving. The two by-products are shorts and bran. *Shorts* represent the intermediate by-product (defined by components) between white flour and bran and consists of small amounts of endosperm, outer layers of the kernel, and the embryo. When the hull is removed, as with the hulless genotype, *bran* is composed almost exclusively of pericarp and testa tissue (Kent and Evers 1994; Posner and Hibbs 1997; Owens 2001). If hulled barley is roller-milled without blocking or pearling, the hull, being cemented to the pericarp and testa tissue, is a part of the bran.

Wheat has been researched, selected, and bred extensively for the most desirable varieties for producing flour through roller milling. Volumes of milling technique data are available to wheat milling companies. Such an array of information does not readily exist for barley milling at this time. Presently there are few or no consistent composition and quality standards available for the fractions derived from the roller milling of barley. The terminology and composition for barley flour and by-products have not been standardized; thus, it is reasonable that terms and composition specifications applied to wheat milling be simply extended to barley milling. As one would expect, barley reacts quite differently from wheat when milled by conventional roller milling techniques to produce flour. Barley bran is more brittle, prone to shatter in rolling, and not readily separated from endosperm, while wheat bran is removed and separated from endosperm tissue in large flakes. Additionally, barley flour flows less readily than does wheat flour. The brittle characteristic of barley bran results in the production of fine particles that increase the ash content and dark coloration of barley flour. Nevertheless, considerable experimental barley milling has been conducted with equipment used routinely in wheat milling, principally to investigate the feasibility of producing barley that is acceptable for baking as well as determining the chemical and nutritional composition of the flour and by-products (Niffenegger 1964; Pomeranz et al. 1971; Cheigh et al. 1975; Sorum 1977; McGuire 1979; Bhatty 1987, 1992, 1993b, 1997, 1999a; Newman and Newman 1991; Wang et al. 1993; Danielson et al. 1996; Xue et al. 1996; Klamczynski and Czuchajowska 1999; Kiryluk et al. 2000).

These early reports demonstrated the practicality of producing barley flour using roller milling equipment and techniques designed for wheat, although many problems were encountered. Wide ranges of flour yields and composition have been reported along with difficulties in adequately separating the flour, shorts, and bran factions. As a result of the early research on the roller milling of barley, it is now recognized that one of the most important factors influencing flour yield is barley genotype. This should not be surprising in view of the well-recognized impact of variety on wheat milling quality. In barley, as opposed to wheat, β-glucans in the endosperm cell wall are of major concern in roller milling. On the one hand, β-glucans are a desirable constituent because of their health-promoting benefits, but on the other hand, they create problems in processing. Waxy and high-amylose barleys generally contain higher levels of β-glucans than do normal starch types, and to a lesser extent, hulless types contain higher levels of β-glucan than do hulled barleys (Xue et al. 1997). The interrelationship of starch type (amylose/amylopectin ratio) and β-glucan level requires different specifications in the milling process to produce maximum yields and high-quality flour. An obvious concern with hulled barley is the inedible hull, but removal of the hull may be accomplished easily by blocking, pearling, or using hulless genotypes, although this creates different milling specifications as well. Blocking or pearling requires additional processing; and hulless barley, regardless of starch type, is reported to require higher power consumption during roller milling, due to resistance in the breakdown of endosperm cell walls (Izydorczyk et al. 2003). If hulled genotypes are roller-milled, the hulls become a major part of the bran and contaminate the shorts fraction as well, posing problems in the use of these by-products.

Starch-type β-glucan interaction was clearly demonstrated by Bhatty (1992), who studied 15 diverse barleys and their roller-milled products, flour and bran. The barleys included in the study were hulled, hulless, and waxy hulless genotypes. After considering an average milling loss of 7%, flour and bran yields were reported as 70 and 30%, respectively. β-Glucan varied from 4.2 to 11.3% in whole-grain barley, 3.9 to 9.0% in flour, and 4.9 to 15.4% in bran. β-Glucan levels in grain, flour, and bran were lowest in hulled (normal starch), intermediate in hulless (normal starch), and highest in hulless waxy starch types (Table 5.3). The bran, which included the shorts fraction, contained 39 to 50% starch, indicating a relatively high level of endosperm tissue remaining in this fraction.

By making certain procedural adjustments, Kiryluk et al. (2000) demonstrated that new and nutritionally valuable products can be produced by roller milling with equipment designed for wheat. Fine- and coarse-grained flours, middlings, and finegrits were obtained from barley tempered to 14% moisture and roller-milled under commercial milling conditions. The fine grits contained about 7, 25, 2, and 16% of β-glucan, total dietary fiber, ash, and protein, respectively, representing significant increases, (about 50, 72, 55, and 24%) in these components over the levels in the original dehulled barley.

Izydorczyk et al. (2003) conducted a study to determine optimum roller-milling conditions for hulless barley that had a wide range of amylose content (0 to 40%) and β-glucan levels (3.9 to 8.1%). Two cultivars in each of four starch types were included in the study, for a total of eight barleys. The four starch types compared had amylose/amylopectin ratios of 25 : 75 (normal), 5 : 95 (waxy), 0 : 100 (waxy), and 40 : 60 (high amylose). A straight-grade flour yield of 74% from normal starch hulless barley was obtained using a milling flow based on traditional wheat

TABLE 5.3 β-Glucan Content of Whole Kernel, Flour, and Bran Produced by Roller Milling Three Diverse Barley Types (g/100 g), Dry Matter Basis

Barley Type	Whole Kernel	Flour[a]	Bran[b]
Hulled ($n = 4$)			
Mean	4.4	4.0	5.1
Range	4.2–4.5	3.9–4.0	4.9–5.4
Hulless ($n = 6$)			
Mean	5.1	4.3	7.1
Range	4.5–5.6	3.8–5.0	6.3–8.1
Hulless waxy ($n = 5$)			
Mean	8.4	6.9	12.3
Range	7.6–11.3	6.3–9.0	10.2–15.4

Source: Bhatty (1992).
[a] 70% yield.
[b] 30% yield.

milling procedures. However, when waxy and high-amylose cultivars were milled using the same flow, the flour yields were less than satisfactory. Modifications of the flow were made that improved the flour yields of these barleys but still not achieving the original 74% yield of the normal-starch barley. Lower flour yields of the waxy and high-amylose cultivars, which had higher levels of β-glucans, confirmed the findings of Bhatty (1999b) and Sundberg and Åman (1994).

Standard roller milling equipment designed to produce wheat flour may have to be modified along with procedural modifications to maximize the milling performance of barley with different types of starch components. Izydorczyk et al. (2003) postulated that the plasticity of the thick endosperm cell walls of the high-β-glucan barleys increases the difficulty of obtaining a clean separation of starch granules and other endosperm cell wall components using the standard wheat milling techniques of successive break and reduction passages. These authors reported two milling techniques that were deemed successful. The first was basically a modified flow normally used in wheat milling as noted above but considered tedious and time consuming (Figure 5.1). Subsequently, these researchers developed a short milling flow that was acceptable for evaluating the milling qualities of the diverse barleys (Figure 5.2). The short flow produced straight-grade flour yields that were about 2 to 3% lower than the long-flow procedure, which was in part compensated by a about 3 to 6% higher yield of a fiber-rich fraction (shorts). However, the β-glucan content of the shorts in the short-flow operation was less than that obtained in the long-flow procedure.

Pearling rates of 10, 20, 30, and 40% were employed to evaluate the roller milling of pearled barley using the short-milling flow described above (Izydorczyk et al. 2003). Yields of flour and shorts were relatively constant when expressed as a percentage of pearled barley. When expressed on a whole barley basis, yield of flour and shorts declined as pearling level increased. Flour

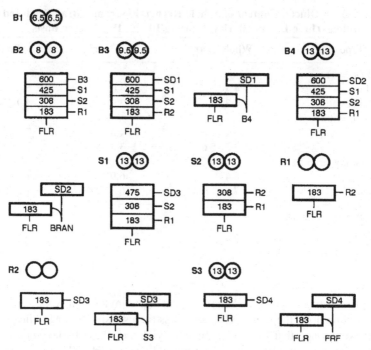

FIGURE 5.1   Long roller mill flow for barley. B, break; R, reduction; S, sizing; SD, shorts duster; FRF, fiber-rich fraction. Numbers on roll symbols indicate corrugations/cm. Numbers on sieves indicate apertures (μm). SD screen has apertures of 140 μm.

brightness was improved up to the 20% rate, but improvement was diminished beyond this point. The normal-starch hulless barley produced the highest flour yield and consumed the least power while producing the lowest shorts yield. High-amylose cultivars consumed the most power and gave the lowest flour yield and highest level of shorts. Pearling prior to roller milling was considered the best compromise between flour yield and fiber-rich fraction yield and flour brightness. Shifts in β-glucan and arabinoxylan concentrations in pearl barley as well as in the major milling fractions were noted as pearling rate increased. Generally, as pearling rate increased, the starch and β-glucan content increased in pearled barley as well as in the flour and shorts, whereas arabinoxylan levels tended to decrease. Pearlings became more enriched in β-glucan as the pearling rate increased up to the 30% rate and then leveled off. Increasing conditioning moisture to 14.5% strongly improved flour brightness, with only a moderate loss of flour yield on a whole unpearled barley basis. Flour yields declined as the conditioning moisture content increased to 16.5%, but the yield of a fiber-rich fraction with greater concentration of β-glucans was increased. It was suggested that when the primary goal is to achieve a maximum yield of a fiber-rich fraction high in β-glucan, conditioning to relatively high moisture levels is indicated.

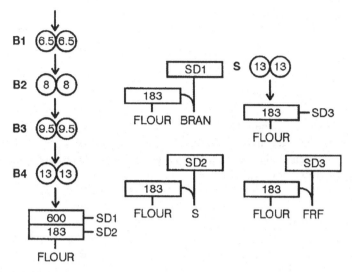

FIGURE 5.2   Short roller mill flow for barley. B, break; R, reduction; S, sizing; SD, shorts duster; FRF, fiber-rich fraction. Numbers on roll symbols indicate corrugations/cm. Numbers on sieves indicate apertures (μm). SD screen has apertures of 140 μm.

The composition and roller milling characteristics of four normal-starch Swedish hulless barleys were reported by Andersson et al. (2003). Two of the barleys were of the same genotype grown at different locations and the others were not closely related other than being hulless. The barleys were roller-milled using a Bühler MLU 202 laboratory mill. Conditioning to 13% moisture was chosen after preliminary studies showed that this produced white flour with ash levels ≤ 0.9%, comparable to the ash content of commercial wheat flour. Flour yield was greater with no conditioning of the barley (11.3% moisture), but the ash content of the flour produced under these conditions was 0.8 to 1.3%, indicating increased amounts of outer kernel tissue in the flour. Further increases in moisture beyond 14.3% resulted in further decreases in flour yield. Average flour, shorts, and bran yields of the four barleys were 43, 51, and 6%, respectively. Yields of the three fractions varied considerably among the unrelated barleys, with less variation between the two samples of the same cultivar grown at two different locations. These observations indicate the possibility of genetic selection among barleys for improving the yield of straight-run flour, as suggested by McGuire (1979). Straight-run flour yields of 37 to 48% as reported by Andersson et al. (2003) are significantly lower than those seen with typical bread wheat. Reduced flour yield was reflected in the higher percentage of the shorts fraction. It is significant to note that in a follow-up baking study by this research group (Trogh et al. 2004), one of the hulless barleys milled in the earlier study (Andersson et al. 2003) was milled industrially (Cerealia, Malmö, Sweden) into flour with a milling yield of 63.7% and protein and ash contents of 13.2 and 1.33%, respectively. Thus, it is possible to obtain somewhat normal flour yields by the roller milling of hulless barley comparable to wheat, but the trade-off is higher percentages of ash, indicating a higher level of bran particles in the flour, which will influence color and perhaps baking characteristics. Andersson et al. (2003) measured the color of the barley flour using grade color factor GCF (units) compared to wheat flour. The GCF values for straight-run flours of the four barleys ranged from 4 to 6 whereas values of about 1 to 3 are typical of wheat flour milled in the same manner. The authors reported that visually the barley flours appeared to be brighter and whiter than wheat flour. However, when the barley flours were moistened, they appeared much grayer than wheat flour, possibly due to activation of enzyme systems.

The chemical composition of the whole barleys (Andersson et al. 2003) was not unusual other than for the consistency of the data among cultivars (Table 5.4). The major fiber components of hulless barley, β-glucan and arabinoxylan, were typical of nonwaxy, hulless Swedish barley grown without environmental stress, as indicated by the average contents of protein (10.8 to 12.5%) and starch (59.2 to 64.5%). As noted previously, the ash content of the barley flour was approximately that expected in milled straight-grade wheat flour, being lower than that of the shorts and bran, with the latter containing the highest level of ash. Protein was lowest in the flour and similar in the shorts and bran fractions. Starch content was highest in the flour, since this was derived from endosperm tissue, but was also relatively high in both shorts and bran, indicating considerable amounts of endosperm tissue in these fractions also. The β-glucan and arabinoxylan levels were lowest in the flour, confirming the findings of Izydorczyk (2003). Shorts

TABLE 5.4   Protein, Starch, Ash, β-Glucan, and Arabinoxylan Content of Four Swedish Whole Hulless Barley and Flour, Shorts, and Bran Produced by Roller Milling (g/100 g), Dry Matter Basis

Item	Protein	Starch	Ash	β-Glucan	Arabinoxylan
Whole grain					
Mean	11.7	61.2	2.1	3.8	4.8
Range	10.8–12.5	59.2–64.6	1.9–2.4	3.0–4.4	4.4–5.3
Flour[a]					
Mean	8.6	77.1	0.9	1.9	1.4
Range	7.7–9.4	75.8–79.3	0.8–1.0	1.6–2.0	1.2–1.5
Shorts					
Mean	14.0	51.5	2.9	5.1	7.1
Range	12.7–14.9	48.2–54.8	2.3–3.3	4.2–5.8	6.1–7.7
Bran					
Mean	14.6	39.3	3.6	3.9	10.2
Range	12.2–15.9	34.8–43.9	2.4–4.5	3.0–4.7	8.1–11.8

Source: Andersson et al. (2003).

[a] Average flour yield = 42.0% (range 36.9 to 47.8%).

contained high levels of β-glucan and an intermediate level of arabinoxylan, whereas bran exhibited the opposite ratio of β-glucan and arabinoxylan, which is representative of the outer layers of the kernel.

Andersson et al. (2003) presented data on the tocol content of the milling fractions. The tocol levels in the whole barleys are shown in Table 4.4. Analyses of the milling fractions showed the highest levels of each of the eight tocol isomers in bran, intermediate in shorts, and least in flour. Tocotrienols (T3) were significantly higher than tocopherols (T) in each milling fraction. Totals of T3 and T (mg/kg) in flour, shorts, and bran were 9.6, 6.0; 45.3, 13.1; and 55.7, 14.0, respectively. The ratios of T3 to T in flour, bran, and shorts were 1.6 : 1, 3.5 : 1, and 4.0 : 1, respectively.

Other research centers have evaluated the nutritional and functional characteristics of hulless barley. Ames et al. (2006) milled 10 hulless barley genotypes in a Bühler mill producing straight-grade flour, shorts, and bran. Composition of the straight-grade flour and shorts and bran fractions exhibited wide genotypic variation. Seven of the genotypes had straight-grade flours with low levels of β-glucan (<3.9%) and dietary fiber (<13%) and high levels of starch (>66%), while straight-grade flour of two genotypes contained >6.5% β-glucan, >18% dietary fiber, and less than 60% starch. The shorts and bran fractions of all genotypes contained higher concentrations of β-glucan and total dietary fiber than those of the straight-grade flour. The shorts and bran fractions from the two genotypes with the high levels of β-glucan and dietary fiber were also higher in these components than in comparable fractions from the other genotypes.

Barley β-glucans present major problems to overcome if barley is to be roller-milled successfully to produce straight-grade flour. Starch type also presents challenges, as in addition to having different granule sizes, the waxy and high-amylose types also contain higher-than-normal levels of β-glucans. Shorts and bran fractions offer possibilities of use to enhance the nutritional components in certain types of baked goods, such as cookies, muffins, cakes, and specialty breads. Major improvements in roller-milling techniques have been made to produce flour from barley that can be used in a multitude of food products.

Malting

Barley has been malted, or germinated, prior to consumption for thousands of years. It has been documented that any barley having a sound, viable kernel will produce malt, but quality factors would be sacrificed in most cases. *Commercial malt* is defined as a modified kernel from which the dried rootlets and sprouts have been separated and removed. In modern malting operations, strict criteria are observed in the selection of barley for malting, not the least of which are the desires and wishes of the brewery, distillery, and/or the brewmaster. A number of major considerations paramount in the choice of a barley for malting include genotypes, kernel size, soundness, color, brightness, a germinating capacity of ≥ 96%, relatively low protein (≤ 12.0%), and the absence of insect, microbial, heat, and weather damage. Observations on the rate of modification of various barleys were noted by maltsters and brewmasters for many years preceding the scientific understanding of modification that is now employed in the malting and brewing industries. In the early years of commercial malting, barley cultivars were selected for superiority in rate and level of modification, with the fewest problems in brewing. Such barleys were deemed *malting cultivars* and the others were termed *feed barleys*, primarily because they were unsuitable for making malt. These cultivars were low-β-glucan barleys having thinner endosperm cell walls and thus could be modified more rapidly with fewer problems in brewing. Barleys have been bred and selected specifically to produce high-yielding superior malts that provide specific objective and subjective characteristics to beverages and foods. After the type of cultivar, a consistently low protein level is perhaps the single major consideration in selecting malting barleys, which is primarily dependent on genotype and environmental growing conditions. Relatively low protein is important to malt production principally because of the negative relationship between the total nitrogen content and the starch content of the kernel, the latter being essential for a high malt extract percentage. On the other hand, there is a positive relationship between total nitrogen content and diastatic power, also a necessary component of malt. *Diastatic power* represents the combined activity of four native enzymes: α-amylase, β-amylase, α-glucosidase, and limit dextrinase, which are essential for the conversion of starch to soluble carbohydrates (malt extract) (Briggs 1978; Burger and LaBerge 1985; Bamforth and Barclay 1993; Kent and Evers 1994; Shewry and Darlington 2003).

The biochemistry of the modification or malting of barley is extremely complicated, not completely understood, and more details are involved than can be presented here. The most obvious change occurring during malting is the destruction of aleurone and endosperm cell walls, but numerous other changes occur that are not as evident. Suffice it to say that a number of enzyme systems and hormones are activated for the process by adding moisture and controlling temperature to a level compatible with germination, provided that dormancy is not a problem. Modification of barley begins at the proximal end of the kernel, adjacent to the scutellum. Access of water to inner portions of the kernel is restricted by nonpermeable covering layers, primarily in the pericarp, and thus water enters principally in the embryo region (Briggs 1978). A small amount of water does penetrate the distal end of the kernel (Axcell et al. (1983) and possibly in the ventral region, but is both much less than and slower than at the proximal end. After enzymes are synthesized in the aleurone and perhaps the scutellum, they migrate into the endosperm to effect hydrolysis of cell walls, proteins, and starch, in that order, according to MacLeod et al. (1964). As noted in Chapter 4, starchy endosperm cell walls are comprised of about 75, 20, and 5% β-glucans, arabinoxylans, and protein, respectively, together with small amounts of cellulose, mannan, and low-molecular-weight phenolics such as ferulic acid (Fincher and Stone 1986). Enzyme systems that are involved in the breakdown of the cell wall structure in the aleurone layer and subsequent hydrolysis of endosperm and embryo nutrients are endo-(1 → 3),(1 → 4)-β-glucanase, endo-(1 → 3)-β-glucanase, endo-(1 → 4)-β-xylanase, endopeptidase, carboxypeptidase, α-amylases (types I, II, and III), β-amylase, α-glucosidase, lipase, phospholipase, and limit dextrinase. There are a host of other enzymes that become active in the germination process, such as superoxide dimutase and catalase, that eliminate peroxide by reacting it with other compounds, such as polyphenols. A range of enzymes (nucleases, phosphatases, nucleotidases, and nucleoside deaminases) are possibly involved in influencing flavor-enhancing nucleotides. Phytase regulates phosphate release from phytic acid and various isoforms of β-galactosidases. Additionally, enzymes present in the aleurone, such as endochitinase, protect the kernel against invasive microorganisms. Endo-β-glucanase, which specifically acts on sections of β-glucan molecules with contiguous β-$(1 \to 3)$ linkages between the glucosyl moieties in the cell walls, is also thought to have a role in protecting against invasive microflora. Gibberellins (plant hormones) are produced in the embryo with the influx of moisture, stimulating enzyme synthesis and activity in the aleurone tissue (MacLeod et al. 1964).

It would seem that these extremely active and diverse enzyme systems would be entirely effective in modifying barley to the smallest intact carbon molecule or mineral component. However, the β-glucans, which are major components of the endosperm cell walls, present problems in modification, filtration, and brewing, a fact that has been recognized for many years (Bamforth 1982; Bamforth and Barclay 1993). It was suggested that the rate of modification of different samples of barley is determined principally by their content of β-glucan and their ability to synthesize β-glucanase. High rates of modification correlate with low β-glucan and high β-glucanase levels (Henry 1989). Conversely, Masak and Basarova (1991) concluded that barleys which are less readily modified develop levels of β-glucanase similar to those in readily modified barleys, but that the β-glucans in the former break down more slowly.

The two main categories of malt are standard malt, which has high diastatic power and a good supply of complex carbohydrates needed for alcoholic beverage production, and specialty malt, which is dried at higher temperatures for a longer time to develop unique flavors and color. Specialty malts are usually intended for uses in food products (Briess Industries, Inc., Chilton, Wiscousin). The most extensive use for barley malt is for the first category, a source of fermentable sugars for alcoholic beverage fermentations. Malt contributes to the flavor and color that are uniquely characteristic of beverages, so much so that a specific beverage product is dependent on the malt for these characteristics to a large degree. Indeed, individual brewers, both large and small, are quite specific in their demands for barley cultivars and conditions during malting to assure consistency in their products. Food or specialty malts are used in lesser quantities, but are added to a number of foods in small amounts, including cooked breakfast cereals, nonalcoholic beverages, sausages, cake mixes, cookies, and other baked foods. Barley cultivars and malting conditions are tightly controlled to produce malts meeting color, flavor, and individual use requirements.

The three basic steps in the malting process are steeping, germination, and kilning (drying). One of barley's advantages over other cereal grains for malting is the presence of an adhering husk to protect the developing acrospires (sprouts) and to aid in filtration of the malt liquor (wort) from the insoluble components (Kent and Evers 1994). In addition, barley has a high starch/protein ratio, a complete enzyme system, self-adjusting pH, desirable color, a neutral flavor, and is economical (Briess Industries). Traditionally, malt has been made from hulled barley, for reasons pointed out above, but Bhatty (1996) investigated malting of hulless barley for food malt applications. The hulless barleys had a shorter steep time and produced malts that were comparable in composition and enzyme activity to hulled barley malts. The hulless barley required a different management technique to prevent damage to emerging rootlets and sprouts, resulting in reduced modification or even death of the kernel. Although hulls are beneficial in preventing sprout damage and acting as a filtering mechanism, once the malting process is complete, hulls become a liability having little nutritive value, even for livestock.

Steeping, the initial process in making malt, starts with raw barley that has been sorted and cleaned and placed into stainless steel tanks with sufficient clean water to wet the grains thoroughly. Water temperature in the steeping process is maintained at 14 to 18°C. The kernel adsorbs water, swells, and starts the initial growth process, wherein multiple enzyme systems act in sequence. Several additives may be used during steeping or during transfer to germination bins to improve and enhance the malting process. These include hypochlorite (calcium or sodium hydroxide, sodium carbonate), formaldehyde, hydrogen peroxide, gibberellic acid, potassium bromate, and ammonium persulfate. Steeping continues for up to 48 to 52 hours, during which time the grain is subjected to alternate wetting and draining, to achieve a moisture content of up to 42 to 46%. Barley will not germinate below 30% moisture and some cultivars require a final moisture level approaching 50%. The activated enzymes begin to break down the protein and carbohydrate matrix (modification) of the endosperm. This exposes the starch and allows development of hormones that initiate growth of the sprouts, sometimes termed *chits*. The barley is commonly transferred to a germination unit (container bin or room) by gravity through a chute, or mechanically by conveyor, or by pumping a mixture of barley and water.

Germination, the second phase of the malting process, begins in the steel tanks, and is continued under controlled conditions of humidity and temperature. Modification of barley starts at the proximal end of the grain adjacent to the scutellum, where the bulk of the moisture enters the kernel. Rate of modification depends on the following: rate of moisture distribution in the kernel, rate of hydrolytic enzyme synthesis, extent of release of these enzymes in the endosperm, and structural features (cell walls) in the endosperm that impede enzymatic hydrolysis. Gibberellin production is at its maximum after the first two days of germination, stimulating the synthesis of the hydrolytic enzymes.

The germinating grain is typically "turned" every 8 to 12 hours to aerate and prevent compaction and "felting," or intertwining of the rootlets. Heat is a product of respiration during germination, requiring its dissipation through aspiration. Air is also required to provide oxygen necessary for metabolism and to remove high concentrations of carbon dioxide, which inhibits the process. The barley kernel is thus modified over a period of four to five days and sometimes up to 10 days in traditional floor malting, which is performed at 13 to 16°C. In modern pneumatic malting houses, temperature is controlled typically over the range 16 to 20°C. Increasing the temperature to 25°C produces roots and enzymes more quickly; however, the rate of enzyme formation decreases such that grains germinating at lower temperatures ultimately contain higher enzymatic activity. Long, cool germination cycles maximize fermentability and also minimize malting loss (Bathgate et al. 1978). Native amylolytic and proteolytic enzymes in the kernel are activated by the moisture and temperature and enhanced by gibberellin, converting insoluble starch into soluble sugars and oligosaccharides. Proteins are converted to free amino acids and peptides by proteolysis. When this modification has reached the maltster's desired stage, the *greenmalt*, as the material is now known, is transferred to the kiln for drying, the third and final stage of processing.

Germination is arrested in the kiln, where warm air passes through the malted grain. The drying process causes withering and stabilization of the grain structure. Kiln drying is divided into four merging phases: free drying down to about 23% moisture, an intermediate drying stage to about 12% moisture, decreasing moisture from 12% to about 6%, and curing where moisture is taken down to 2 to 3%. During free drying, an air temperature of 50 to 60°C is attained along with an airflow of 5000 to 6000 m^3/min, causing unrestricted removal of water. When moisture content reaches about 12%, the water in the malted kernel is said to be bound, requiring further increase in air temperatures and reduction of air fan speed. Lowering of moisture to about 6% allows the malt to be "cured" with the air temperature increased to 80 to 110°C, reducing moisture to 2 to 3%. Drying time and temperatures are essential in developing the flavor and color characteristics of the finished malt, this being a critical point in the malting process. Making specialty malts, as for use in food, differs from the basic malting process for beverage production in that batch sizes are smaller and constant monitoring is required to meet specifications (Kent and Evers 1994; Gibson 2001; Briess Industries). Malted kernels can be used directly in breads or other bakery products or processed further into flour extracts or syrup. A characteristic of malt products is the presence or absence of enzymes. Malt ingredients that contain diastatic enzymes are used in brewing. Nondiastatic malt products are processed at varying higher temperatures, which destroys some or all of the enzymatic activity. Diastatic malt flour is commonly used in bread baking in small amounts to act as a dough conditioner (Hansen and Wasdovitch 2005).

Malt employed in the distilling industry to produce whiskey is made from maize, rye, and barley grains. Specifications placed on malt for distilling are comparable to those employed for brewing malt. Residual levels of unfermented dextrins are irrelevant to the flavor of whiskey, although they are desirable in beer and food malts. The goal is the maximum production of ethanol. Malt vinegars are produced by the bacterial oxidation of alcohol from wine, cider, or beer to acetic acid. Production of malt vinegars requires high fermentability of the grist so as to maximize ethanol production, as in distilling. Accordingly, high-diastatic malt may be employed that is capable of maximum conversion of starch in the adjunct. Acetic acid bacteria such as *Acetobacter aceti* are used to oxidize ethanol to acetic acid. Specifications applied to malt for vinegar manufacture relate to yield of soluble material and yield of fermentable material. The use of malt-based ingredients in foods is discussed briefly in Chapter 7.

The effect of modification on nutrient composition in hulless and standard malting barley is described in Table 5.5 (Bhatty 1996). The absolute values shown for protein and starch are not indicative of the substantial enzymatic hydrolysis that occurs to these nutrients. Bhatty (1996) suggested that this abnormality was due to the analytical procedures used to measure these components. The average increases shown in soluble carbohydrates (3.5 to 10.3%, soluble protein (2.8 to 4.4%), and reducing sugars (0.3 to 3.1%) and decreases in β-glucans (4.7 to 1.5%) and viscosity (26.8 to 1.7 cP) present a truer indication of the hydrolytic activity occurring during modification. The soluble/total protein (S/T) ratio, known as the *Kolbach index*, is a measure of proteolysis. The higher the S/T ratio, the more extensive is the breakdown of protein. The lower β-glucan content was largely responsible for the greatly reduced viscosity. The major increases seen in α-amylase (ceralpha units), diastatic power, β-glucanase, and proteolytic activity are directly responsible for the changes. In comparing the hulless barley to the standard hulled malting barley, the major composition differences noted were in higher levels of α-amylase and diastatic power in the standard barley, but higher levels of β-glucanase in the hulless barley. Proteolytic activity was not greatly different between the two barley genotypes. Caution should be exercised in this area, due to the limited data available.

Lipid constituents and their influence on the malting process and malt have not been investigated as extensively as have carbohydrates and protein. Kaukovirta-Norja et al. (1997) reported total lipids, including internal starch lipids, to be about 3.7% in the intact kernel prior to malting and about 3.3% in the resulting

TABLE 5.5   Effect of Malting on the Composition, Viscosity, and Enzymatic Activity of Hulled and Hulless Barley, Dry Matter Basis

Item	Hulled Barley		Hulless Barley	
	Unmalted	Malted	Unmalted	Malted
Protein (%)	15.3	13.7	16.0	15.7
Ash (%)	2.4	2.2	1.8	1.4
Starch (%)	61.4	62.8	64.4	62.9
β-Glucan (%)	4.4	1.2	4.9	1.8
Viscosity (cP)	20.2	1.6	33.3	1.7
α-Amylase (ceralpha units/g)	0.1	258.4	0.1	288.0
Diastic power (°ASBC)	71.0	189.5	113.2	147.5
β-Glucanase (U/kg)	0	583.0	0	502.0
Proteolytic activity (mg NPN/kg)	42.0	112.0	46.0	105.0

Source: Bhatty (1996).

malt. These levels are 16 to 18% higher than reported in comparable barley and barley malt, respectively where the starch lipids are not included. Shifts in chemical structure occur in surface and internal lipids associated with starch granules during malting, which suggests that such lipid changes may be responsible for oxidation and the formation of off-flavors in some malts.

SECONDARY PROCESSING

Extrusion

Extrusion is a technology used for conversion of basic dense grain formulations into light, puffed, and crisp products that are cooked (RTE) or partially cooked. It is a process using high-temperature short-term cooking applicable to producing RTE breakfast cereal and snack foods and for direct expansion and forming of solid shapes. The basic components and operation of extruders are screw(s) rotating in a barrel propelling food material forward through a die or restrictive orifice, generating heat and pressure in the process. The action of these components on the substance being extruded is influenced directly by the operating pressure in the system. Miller (1990) provided a geometric and physical explanation of the principles of die flow. Raw ingredients are first moved into a cooking chamber, where they are compressed and shredded at elevated temperatures to undergo a melt transition and form a viscous fluid. The extruder develops the fluid by shearing the biopolymers, particularly starch. *Shear* is a term referring to the resistance of molecules as they move around a mixture, due to the presence of intermolecular forces. Starch has a high water-binding capacity during cooking, which causes viscosity development. The die pressure and increased temperature cause reorganization of the food molecular structure. The fluid is then forced through dies to form a variety of shapes with low density (Guy 1991).

Extruders may consist of single or twin screws with spirally arranged flights for conveying material through a barrel or tube with a capacity for developing pressure. Heat is produced by steam injection as well as the mechanical energy developed by the pressure applied. Single-screw extruders were first used to produce RTE breakfast cereals in the 1960s. Problems were related to movement through the barrel and excessive shear, affecting color and appearance of finished products as well as overgelatinization. Twin-screw extruders have greater efficiency and produce increased solubility of the material being extruded, which allows for greater control (Kent and Evers 1994). Fast (2001) described innovations in twin-screw extruders and the advantages in control of the various steps in the extrusion process, including advantages in die design and cutting mechanisms. Extrusion is a rapidly expanding technology for processing cereals into RTE breakfast cereals, snack foods, and similar products. New developments include improved designs in all phases of the equipment that allow for multiple use and greater control of shear, compression, and pressure at the die. The promotion of whole grains in modern dietary recommendations and subsequent health claim labeling for whole grains (FDA 1999) has resulted in new extruded products containing larger concentrations of bran. This presents a set of unique problems to many food manufacturers, particularly in achieving palatability, texture, and mouthfeel in products (Fulcher and Rooney-Duke 2002). Replacing some of the starch in mixtures to be extruded by inclusion of fibrous ingredients alters the plasticity of the mixture and shifts the balance between expandable and nonexpandable material. The results of added fibrous material are products with lower expansion and higher density. Production of such products requires modification of extruder conditions to achieve consumer acceptable products (Fulger 1988).

As in other processing methods, barley presents a unique blend of components that significantly influence extrusion processing characteristics and end products. Primary factors are the β-glucan and TDF contents, but starch type (normal, waxy, and high-amylose) can also require major procedural adjustments to apply extrusion techniques successfully to barley. Although limited information is available on extrusion processing of barley flour, a few reports appear in the scientific literature. Østergård et al. (1989) reported that dietary fiber content of barley flour increased with extrusion cooking, whereas starch content decreased. Heryford (1987) extruded four barley cultivars using a laboratory-scale single-screw extruder (Insta-Pro, Des Moines, Iowa) and reported increased viscosity in all of the products extruded, although total and soluble β-glucans were decreased by extrusion. This effect was probably due to molecular changes in the starch. Berglund et al. (1994) studied the feasibility of producing RTE breakfast cereals by extruding barley flour and blends of barley flour with rice or wheat flour using a corotating twin-screw extruder. Cereals produced by extrusion of 100% barley flour had limited expansion and high bulk density. Blending barley flour with 50% rice flour reduced bulk density by 50%, producing a product similar in appearance to 100% rice flour cereal. Extrusion increased β-glucan content and extract viscosity measurements of the barley flour blends. These authors concluded that extruding barley flour blends with rice flour produced RTE cereals comparable in crispness, color, and flavor to that of extruded 100% rice flour products.

Huth et al. (2000) studied the effects of extrusion on the formation of resistant starch in two barley cultivars differing in amylose content. The varieties studied were Korna, containing 22.4% amylose starch and 2.4% total β-glucan, and HiAmi Cheri having starch that contained 29.0% amylose and 6.1% total β-glucan. After dehulling and pin-milling the barley, extrusion was accomplished with a single-screw laboratory extruder (DN 20, Brabender, Duisburg, Germany). The highest contents of resistant starch (6%) were generated by using a mass temperature of about 150°C and about 20% moisture during extrusion followed by storage at −18°C for 3 to 7 days. Under these conditions, the macromolecular state of β-glucan was preserved. These authors concluded that barley cultivars with high amylose and high β-glucan contents are preferred to produce extruded barley foods with enhanced nutritional qualities.

The effects of extrusion cooking on properties of barley β-glucans in nonwaxy and waxy barley cultivars were examined by Gaosong and Vasanthan (2000). The barleys were pearled and pin-milled into flour prior to extrusion. For both cultivars, β-glucan in the extruded flours had solubility values higher than those of the raw flours. Purified β-glucan was extracted from the raw and extruded flours, and the solubility of these samples was found to differ between cultivars and extrusion conditions. These researchers also examined the glyco-sidic linkage profiles in an attempt to explain solubility differences and were unable to arrive at a satisfactory conclusion. Baik et al. (2003) extruded break and reduction flours from two normal starch (nonwaxy) and two waxy starch barley cultivars. The barley flours were formed into granules before extrusion. Break stream flours, generally with higher starch content than reduction flours, produced extrudates with higher expansion indexes. Nonwaxy starch barleys had higher expansion indexes and lower density values than those of waxy barley. These researchers concluded that granulation of the barley flour and fine control of extrusion conditions were essential factors in obtaining an acceptable RTE breakfast cereal.

The effects of extrusion cooking on the dietary fiber profiles of waxy and nonwaxy barley flours were reported by Vasanthan et al. (2002). Flour was pre-pared by pearling the grain in a laboratory pearler (Satake model TM05, Tokyo, Japan), removing 30 to 32% of the kernel, followed by pin milling (Alpine Con-traplex wide chamber mill, type A-250, Hosokawa Micron Systems, Summit, New Jersey). The flours were extruded in a twin-screw extruder (Werner and Pfleiderer ZSK 57 W 50P, Stuttgart, Germany). The content of soluble dietary fiber and total dietary fiber increased upon extrusion cooking of both waxy and nonwaxy barley flours. Changes in insoluble dietary fiber were observed to be cultivar dependent. Insoluble dietary fiber was increased in the nonwaxy flour while decreasing slightly in the waxy flour, due to extrusion. It was suggested that the formation of retrograded amylose caused the formation of some resis-tant starch in the nonwaxy flour, thus increasing the level of insoluble dietary fiber. Increases in soluble dietary fiber in both types of extruded barley products were thought to be due to the transformation of some insoluble dietary fiber into soluble dietary fiber and the formation of additional soluble dietary fiber by transglycosidation.

Through the use of various types of extrusion equipment and techniques along with select barley cultivars, food products with new and enhanced functional and nutritional properties may be produced to meet consumer needs and demands. Extrusion technology that is currently available to the modern food industry can be used to produce many useful products that are beneficial to the health of consumers.

Steel Cutting

Steel-cut barley or barley grits are made with rotary granulators or cutters, con-sisting of a revolving drum, perforated with round countersunk holes through which the kernels can flow out after cutting. A series of adjustable knives are mounted at the lower side of the drum. Depending on the angle of the knives, which are adjustable, the kernels can be cut into varying sizes. A sifter is used to remove the fines and, if desired, to classify the cut kernels into specific sizes and flourlike material. Uncut barley is separated from the cut barley in the screening and recycled to the cutter (Vorwerck 1992). If the germ is included in cut barley products, analytical values are similar to the whole or blocked barley, or if pearled barley is cut, the values would be comparable to those of the original pearled barley.

Flaking

The process of making flaked barley is similar to that of producing oat flakes or rolled oats from dehulled oats or oat groats. Flaking produces the desired change in the texture for easier preparation (shorter cooking time) and improved digestibility. Depending on the desired size and thickness of the flakes, the initial product will be either blocked, pearled, or steel cut. Barley thus processed can be compared to oat groats in the flaking process, although some differences in cutting, shattering, and moisture absorption between oat groats and barley would be expected as well as differences occurring between barleys, especially between cultivars having different starch types. The grains are first cleaned and sized and then conditioned (tempered) by increasing the moisture content by 2 to 4%, then passed through a kiln, where they are heated to a temperature of 99 to 104°C. Heating deactivates undesirable enzymes in the raw grain, produces a roasted flavor, increases elasticity, and gelatinizes the starch. The conditioned grain may be steel cut and separated into small and large pieces by sieving. Flaking is accomplished by passing hot, moist grains through flaking rollers, which flatten the kernel, producing flakes of various thicknesses. Flaking rolls are heavier and larger than flour milling rolls and can be adjusted to control a desired thickness of the flakes. Instant or quick-cooking flakes are about 0.25 to 0.38 mm thick, whereas whole traditional or "old-fashioned" flakes may be 0.5 to 0.76 mm thick. After rolling, the flakes are cooled and dried before storage or packaging (Caldwell et al. 1990; Vorwerck 1992; Kent and Evers 1994).

Infrared Processing

Infrared processing is an alternative heating technique described as thermal radiation. When a refractory surface such as metal or ceramic is heated to a certain temperature, it emits electromagnetic waves within the infrared spectrum. The wavelength in the infrared spectrum ranges from 0.76 μm to 1.0 mm, although the infrared band in which infrared heating with which is mostly concerned ranges from 0.76 μm to a maximum of 400 μm. Short or near-infrared wavelength ranges from 0.76 to 2.0 μm, and medium wavelengths range from 2.0 to 4.0 μm. The wavelengths created within the infrared spectrum are dependent on the temperature of the refractor surface. The most common wavelengths employed in infrared heaters have an energy distribution and wavelength typically centered around 2.7 μm. This is considered a medium infrared wavelength and corresponds to the spectrum in which water and many other materials effectively absorb heat energy. The basic concepts of infrared radiation are high heat-transfer capacity, heat penetration directly into the product, fast regulation response, and no heating of surrounding air in the oven.

Infrared ovens are designed with gas-fired (open flame) or electric heaters (calrods) that heat the refractor surface. The flame or calrods are not sources of infrared radiation but function to heat the refractor surface to temperatures ranging from 900 to 1100°C (1650 to 2000°F). The heated refractor surface then emits electromagnetic waves or infrared radiation, which is directed to the material (food), where it is absorbed, transmitted, or reflected. As the temperature of the refractor surface is increased, the maximum radiation occurs at shorter wavelengths and has a much higher intensity, with an increasingly larger portion of the radiation occurring in the near-infrared spectrum. The radiation absorbed causes the molecules within the food to vibrate in accordance with the resonance of their own wavelength frequency. This produces rapid internal frictional heating and a rise in water vapor pressure. In the case of a cereal grain such as barley, infrared heating causes the starch granules to swell, fracture, and gelatinize, permitting the grain to be pearled, ground, or flaked without conventional conditioning. Infrared radiation is considered an excellent means of predrying "sheeted" products such as taco chips, tortillas, and masa. The thin dough used in these products dries out completely and quickly, preparing them for further processing, such as deep-fat frying or further toasting by conventional or infrared heating. The short wavelengths of infrared radiation commonly employed do not penetrate deeply into thicker dough, such as in pocket breads. These breads become layered, ending up with a tightly sealed "skin" on the outside, with lighter, fluffier material inside, due to the high temperature gradient. Infrared heat processing has applications in quick-cooking food products; however, the most widespread and common uses of infrared heating are for drying, toasting, and browning (Skjöldebrand and Andersson 1987; Anonymous 2007; Red-Ray Manufacturing Company Cliffside Park, New Jersty).

Lawrence (1973) studied the effect of infrared-heating on the composition of ground and flaked hulled barley. Dry matter was increased by 2.4% and nitrogen was reduced by 0.5% without any consistent effect on lipid or crude fiber contents of the barley due to infrared cooking. Newman et al. (1994) reported similar findings in that dry matter was slightly increased (1.0%) and nitrogen was reduced (5.3%) in waxy hulless barley following infrared heating to 900°C for 3 minutes followed by flaking. However, in contrast to the report of Lawrence (1973), fiber content was altered by the infrared heating procedure. Insoluble dietary fiber was decreased by 14.8% (8.1 to 6.9 g/100 g) and soluble dietary fiber was increased by 50.0% (3.0 to 4.5 g/100 g), along with a 3.5% increase in β-glucans (6.28 to 6.48 g/100 g). The increased fiber and β-glucans resulted in a 30.7% increase in extract viscosity (2.96 to 3.87 cP). It should be noted that the fiber reported in the two studies was determined by different procedures.

Ames et al. (2006) presented data on the effect of infrared treatment of four Canadian hulless barley genotypes. The grains were tempered to various moisture levels (16 to 28%) prior to a 70-second infrared treatment. Materials were then evaluated as whole grain and fractions after roller milling. The heat treatment inactivated the peroxidase enzyme, which is indicative of destroying the activity of several enzymes in grain that reduce shelf life and the nutritional quality of flour and by-products. Infrared treatment also had a positive effect on how barley kernels fractionated during milling, resulting in increased yields of straight-grade flour. Roller milling yields of straight-grade flour were increased from 38% for untreated barley to 42, 43, and 46% at moisture conditioning levels of 16, 20, and 24%, respectively. Additionally, β-glucan and TDF levels were increased in the straight-grade flour at each increment of added moisture, accompanied by large increases in extract viscosity. These findings regarding β-glucan and TDF support the data of Newman et al. (1994). Ames and her group also evaluated the effect of micronizing on the extractability of β-glucans from milling fractions for possible development of beverages such as hot barley tea. Significant improvements were found in the extractability of β-glucans from milling fractions produced from micronized barley. The extractability of β-glucan was increased in a fine fraction from 33.6 to 41.6%, and in a coarse fraction from 2.6 to 4.2%, as well as in whole-meal barley from 15.7 to 21.7%.

Infrared processing or micronizing appears to be a promising processing technique for improving yield, functionality, and nutritional quality of barley food products.

SEPARATION TECHNIQUES

Air Classification

Air classification is a separation technique that uses air to effect dry separation of mixtures having certain characteristics into particle size groups ranging from 2 μm to a maximum of 100 μm, referred to as *cut points* or *cut size*. The cut size is defined as the particle size that has a 50–50 chance of ending up in the fine fraction or in the coarse fraction. In the modern food industry, air classification is generally utilized to separate particulates in the range 2 to 60 μm. The size, shape, and density of the particles in the mixture and the physicochemical nature of the product are the characteristics considered to be important for successful air classification. The procedure is an operation that applies the technology of moving air in a confined space to separate nonhomogeneous particles into groups or classes of fairly uniform size based on density or mass of the particles. In general, air classifiers will segregate a heterogeneous mixture into two groups, one primarily below the targeted cut size (fine fraction) and the remainder above this size (coarse fraction). Such fractions may be reclassified separately to produce additional fractions. Air classifiers complement screens in applications requiring cut sizes below commercial screen sizes and supplement sieves and screens for coarser cuts (Fedec 2003). The general principles of air-classifying devices are as follows: The material to be classified is suspended in an airstream, the air–solids stream is introduced into the classification zone, and the coarse fraction is seperated from the fine fraction and airstream by opposing the drag force created by the air with gravitational, inertial, or centrifugal force or a combination. In some applications it is necessary to remove part of the airstream with the coarse fraction to remove the coarse particles, then separate and collect the fine fraction from the airstream. (Fisher-Klosterman, Inc, Louisville, Kentncky).

To apply air classification to barley requires barley to be hulless, blocked, or pearled and prepared as whole-meal flour by impact milling. Pin milling is the usual choice for impact milling prior to air classification because the particle size obtained is more uniform than that using hammer milling. Air classifiers are available that combine impact milling for the necessary fine grind, along with classification. Particle size is important in the air classification of all cereal flours. Particles should be sufficiently small so that cell components can be separated (King and Dietz 1987). The moisture and fat contents of barley flour are important factors to consider prior to air classification. The moisture level should be in the range of 8%, with a fat content 2% or less (Fedec 2003). Air classification can sort the flour selectively into at least three fractions, depending on the cut size specified: protein-rich fines up to 15 to 18 μm, starch-rich fraction in the range 15 to 35 μm, and a coarse fraction containing cell wall material (dietary fiber) from 33 to 45 μm (Jadhav et al. 1998). More fractions may be obtained by regrinding and reclassifying selected fractions.

Over 35 years ago Pomeranz et al. (1971) suggested that air classification would be useful in developing high-protein fractions from milled barley that could be used in foods and high-starch fractions for use as adjuncts by the brewing industry. Vose and Youngs (1978) air-classified malted barley following roller milling, producing flour fractions having differing composition, flavor, aroma, and amylolytic activities. α-Amylase activity was concentrated in the high starch fraction, and the malt aroma and flavor were concentrated in the less-dense, low-fiber, high-protein fraction. Wu et al. (1994) demonstrated that grinding and air classification of barley meal resulted in good yields of fractions having enriched β-glucan and protein, but the cultivar had a significant influence on the end products. These authors showed that air classification of dehulled commercial barley (cv. Portage) resulted in a good protein shift and good β-glucan values. Air classification of a high protein hulless cultivar (Hiproly Normal) and a high protein waxy hulless cultivar (Prowashonupana) yielded low protein shift and high β-glucan values. The combined high β-glucan fractions from Prowashonupana contained 31% β-glucan.

Vasanthan and Bhatty (1995) reported a study on pin milling and air classification of waxy, normal (starch), and high-amylose barleys into coarse and fine fractions. The coarse fractions ranged from15 to 36% starch, 12 to 15% protein, and 13 to 24% β-glucan. The fine fractions ranged from 77 to 78% starch, 8 to 9% protein, and 6 to 8% β-glucans. The shifts in starch, protein, and β-glucan varied with cultivar. Coarse fractions from the waxy and high-amylose cultivars had the highest levels of β-glucan, 24 and 23%, respectively. Seven barleys, representing hulled and hulless barleys with normal, waxy, and high-amylose starches, were air classified after impact grinding in a study reported by Andersson et al. (2000). Enrichment of starch, protein, and β-glucan was accomplished by air classification, but significant variations occurred in the nutrient concentrations due to starch and hull type. The reader is referred to Fedec (2003) for a more detailed description of the principles of operation and uses of air classification in the food industry, including sources for equipment and plant system designs.

Sieving

Sieving is an alternative method to air classification for producing protein- and fiber-rich fractions from pin- or hammer-milled barley (Knuckles et al. 1992; Yoon et al. 1995) and can be combined successfully with air classification (Knuckles and Chiu 1995; Sundberg et al. 1995). The term *sieving* as used in the grain milling and processing industries refers to the separation of ground material into various particle size classifications, utilizing size-designed sieves or screens (Posner and Hibbs 1997). Sieving is also a component of the roller-milling process used to produce high-quality bread flour, but also may be applied to whole-meal flour prepared with other milling machinery, such as hammer mills or pin mills. Separation of ground barley by sieving is more effective with dehulled barley (blocked), pearled or hulless barley, than with hulled barley.

Separation of flour-meal particles in a sieve is accomplished by moving the particles across a sieve screen, resulting in the smaller particles passing through the screen and the larger particles being removed to a collection point by a conveyor. Posner and Hibbs (1997) described the six principles that govern the results obtained through sieving as direction of movement of the sieve (gyrating, reciprocating, a combination of the two, or rotating), rate of movement of the flour relative to sieve surface area, the size of the apertures of the sieve, the amount of sieve surface area, the amount of flour on the sieve surface, and the granulation and shape of the flour particles.

In a commercial wheat flour milling operation, which may be adapted to barley, a sifter section is a box containing a stack of sieves arranged to separate material flowing over and through the sieves into different particle size fractions or streams as determined by the principles outlined above. The sieve arrangement is known as the *flow of the sifter*, and arranging of the sifter sections is called *flowing* the sections. Sifter sections can be flowed to remove coarse material first and the fine material last at the bottom of the sifter stack or arranged in reverse with fine material removed first and coarse material removed last. The sieve arrangement is referred to as *sieve stacking* (Posner and Hibbs 1997).

Air classification and sieving are effective methods to separate materials such as whole-meal barley or barley flour into fractions with different levels of protein, starch, and dietary fiber for application in a variety of food products. A combination of air classification and sieving is often more effective in isolating particles of different sizes and composition than is either procedure alone. Future food and industrial processors may require ingredients for standard as well as specialty products containing a wide variety of physical, chemical, and functional properties, depending on end use. Thus it is not appropriate to describe a particular product or fraction of the milling processes to be superior to another, only that they are different and each product possesses different functional characteristics (Jadhav et al. 1998).

SPECIAL AND MISCELLANEOUS PROCESSES

The location of various nutrients in the barley kernel permits the concentration and semi-isolation of a number of these important nutritional compounds by taking advantage of standard grain processing procedures. Products that contain concentrated levels of various nutrients are produced in pearling and roller milling. For example, phenolic compounds are located in the outer areas of the kernel and are concentrated in the pearlings in the process of making pearl barley (Lampi et al. 2004) and in roller-milling fractions (Andersson et al. 2003). Barley pearlings also contain concentrated levels of other nutrients, such as oil and vitamin E (Wang et al. 1993; Peterson 1994; Marconi et al. 2000). Air classification and sieving are effective methods of producing fractions of ground barley having enhanced levels of different components, primarily protein and β-glucan. Concentration, isolation, and purification of barley β-glucans received considerable attention even before the official recognition by the FDA (2006) of health benefits of β-glucans relating to control and prevention of cardiovascular disease in humans (Klopenstein and Hoseney 1987; Goering and Eslick 1989; Bhatty 1992, 1993a, 1993b; Wu et al. 1994; Knuckles and Chiu 1995; Sundberg et al. 1995; Vasanthan and Bhatty 1995; Andersson et al. 2000). These studies utilized standard dry-milling processing, such as pearling, roller milling, sieving, and air classification, to concentrate the β-glucans. In roller milling and pearling, the outer seed coat is separated from the endosperm, producing products that are representative of the two areas of the kernel that may be refined by air classification and sieving. Air classification separates finely ground grain into small, medium, and large fractions having different particle sizes which differ in nutrient content. Further air classification and/or sieving of each fraction can produce even greater concentrations of specific nutrients. Dietary fiber is associated with larger particles, while the small particles contain higher levels of starch and protein.

Solvent extraction procedures have been researched that produce products with increased levels (70 to 80%) of β-glucan (Goering and Eslick 1989; Bhatty 1993c; Burkas and Temelli 2004). More recently, a relatively pure high-β-glucan (about 70%) product has been produced on a commercial scale from waxy hulless barley for the purpose of supplementing numerous food products (L. Kolberg, personal communication, 2006). A β-glucan concentrate purified from barley fiber with nearly 90% purity was recently reported to have been developed (Zhang et al. 2007).

Linko et al. (1983) and Goering and Eslick (1989) developed processes for the concentration of maltose, producing high-maltose syrup from barley. The patented process of Goering and Eslick (1989) included the production of barley β-glucan concentrate, bran protein, and oil in addition to high-maltose syrup. Cowan and Mollgaard (1988) developed a procedure to prepare hydrolyzed β-glucan that compared favorably with sucrose for use in various foods. Jadhav et al. (1998) summarized the potential industrial and nonconventional food applications of barley starch for the manufacture of paper, textiles, adhesives, adsorbents, binders, flocculants, biodegradable products, detergents, and for chemical or biological conversion to versatile compounds such as levulinic acid and 5-hydroxymethyl-fuyrfural.

SUMMARY

Recognition of the impact of the chemical composition of barley on milling and processing techniques is no less important than selecting the appropriate barley genotype. A thorough knowledge of cereal processing and the impact of barley genotype on such processes is an absolutely necessary requisite to the manufacture of appealing, tasty, and acceptable as well as healthy food products. Although barley has a long history as a food, it must compete with maize, oats, rice, rye, and wheat, which are already well accepted food commodities in the marketplace. At the present time, the general public recognizes barley as a food primarily as pearled barley, an ingredient in soups and stews. Pearled barley is an important food product. Extensive food research has demonstrated the versatility and adaptability of barley for many more foods than just pearled barley. Pearling has been studied extensively, in part due to the adaptability of barley to equipment designed for polishing rice. The advent of new barley cultivars, starch types, and hulless genotypes has presented the cereal food industry an opportunity to develop new pearled products that are exciting and have many health benefits. It has been demonstrated that roller milling, traditionally a process for producing wheat flour, can be adapted to use for producing barley flour. Research has demonstrated that most of the machinery and procedures originally designed for processing wheat, oats, or rice can be adapted to process barley, allowing the production of new cereal products that offer diversity in taste as well as improved health-promoting characteristics. Although malting is considered to be primarily a procedure for preparing barley for alcoholic beverage production, numerous foods benefit from the inclusion of small amounts of food-grade barley malt, including ice cream.

The by-products of pearling and milling have traditionally been utilized in feedstuffs for farm animals and pets. Recognition of the health benefits of parts of the kernel normally considered as by-products makes these products valuable to enhance many foodstuffs, some of which are currently being marketed, and a vast area for health food product development.

REFERENCES

Ames, N., Rhymer, C., Rossnagel, B., Therrien, M., Ryland, D., Dua, S., and Ross, K. 2006. Utilization of diverse hulless barley properties to maximize food product quality. *Cereal Foods World* 51:23–28.

Andersson, A. A. M., Andersson, R., and Åman, P. 2000. Air classification of barley flours. *Cereal Chem.* 77:463–467.

Andersson, A. A. M., Courtin, C. M., Delcour, J. A., Fredriksson, H., Schofield, J. D., Trogh, I., Tsiami, A. A., and Åman, P. 2003. Milling performance of North European hull-less barleys and characterization of resultant millstreams. *Cereal Chem.* 80:667–673.

Anonymous. 2007. *Natural gas proves superior for infra-red food processing.* Industrial Processing, Natural Gas Application in Business and Industry. Published online at http://www.Energysolutionscenter/resources/PDFsGT_S03_Infrared_Food_Processing. pdf.

Axcell, B., Jankovchy, D., and Morrell, P. 1983. Steeping: the crucial factor in determining malt quality. *Brew. Dig.* 58:20–23.

Bach Knudsen, K. E., and Eggum, B. O. 1984. The nutritive value of botanically defined mill fractions of barley: 3. The protein and energy value of pericarp, testa, germ, aleurone and endosperm rich decortication fractions of the variety Bomi. *Z. Tierphysiol. Tierernaehr. Futtermittelkd.* 51:130–148.

Baik, B.-K., Powers, J., and Nguyen, L. T. 2003. Extrusion of regular and waxy barley flours for production of expanded cereals. *Cereal Chem.* 81:94–99.

Bamforth, C. W. 1982. Barley β-glucans: their role in malting and brewing. *Brew. Dig.* 57:22–27.

Bamforth, C. W., and Barclay, A. H. P. 1993. Malting technology and the uses of malt. Pages 297–354 in: *Barley: Chemistry and Technology.* A. W. MacGregor and R. S. Bhatty, eds. American Association Cereal of Chemists, St. Paul, MN.

Bathgate, G. N., Martinez-Frias, J., and Stark, J. R. 1978. Factors controlling the fermentable extract in distillers malt. *J. Inst. Brew.* 84:22–29.

Berglund, P. T., Fastnaught, C. E., and Holm, E. T. 1994. Physicochemical and sensory evaluation of extruded high fiber barley cereals. *Cereal Chem.* 71:91–95.

Bhatty, R. S. 1987. Milling yield and flour quality of hull-less barley. *Cereal Foods World* 32:268–272.

Bhatty, R. S. 1992. β-Glucan content and viscosities of barleys and their roller-milled flour and bran products. *Cereal Chem.* 69:469–471.

Bhatty, R. S. 1993a. Nonmalting uses of barley. Pages 355–417 in: *Barley: Chemistry and Technology.* A. W. MacGregor and R. S. Bhatty, eds. American Association Cereal of Chemists, St. Paul, MN.

Bhatty, R. S. 1993b. Physicochemical properties of roller-milled barley bran and flour *Cereal Chem.* 70:397–402.

Bhatty, R. S. 1993c. Extraction and enrichment of (1–3) (1–4)-β-D-glucan from barley and oat brans. *Cereal Chem.* 70:73–77.

Bhatty, R. S. 1996. Production of food malt from hulless barley. *Cereal Chem.* 73:75–80.

Bhatty, R. S. 1997. Milling of regular and waxy starch hull-less barley for the production of bran and flour. *Cereal Chem.* 74:693–699.

Bhatty, R. S. 1999a. β-Glucan and flour yield of hull-less barley. *Cereal Chem.* 76:314–315.

Bhatty, R. S. 1999b. The potential of hull-less barley. *Cereal Chem.* 76:589–599.

Bhatty, R. S., and Rossnagel, B. G. 1998. Comparison of pearled and unpearled Canadian and Japanese barleys. *Cereal Chem.* 75:15–21.

Briggs, D. E. 1978. *Barley*. Chapman & Hall, London.

Burger, W. C., and LaBerge, D. E. 1985. Malting and brewing quality. Pages 367–401 in: *Barley*. Agronomy Monograph 26. D. C. Rasmussen, ed. American Society of Agronomy Crop Science Society of America, and Soil Science Soceity of America, Madison, WI.

Burkus, Z., and Temelli, F. 2004. Rheological properties of barley β-glucan. *Carbohydr. Polym.* 59:459–565.

Caldwell, E. F., Dahl, M., Fast, R. B., and Seibert, S. E. 1990. Hot cereals. Pages 257–259 in: *Breakfast Cereals and How They are Made*. R. B. Fast and E. F. Caldwell, eds. American Association Cereal of Chemists, St. Paul, MN.

Cheigh, H. S., Snyder, H. E., and Kwon, T. W. 1975. Rheological and milling characteristics of naked and covered barley varieties. *Korean J. Food Sci. Technol.* 7:85–95.

Cowan, W. D., and Mollgaard, A. 1988. Alternative uses of barley components in the food and feed industries. Pages 35–41 in: *Alternative End Uses of Barley*. D. B. H. Sparrow, R. C. M. Lance, and R. J. Henry, eds. Waite Agricultural Research Institute, Glen Osmond, Australia.

Danielson, A. D., Newman, R. K., and Newman, C. W. 1996. Proximate analysis, β-glucan, fiber and viscosity of select barley milling fractions. *Cereal Res. Commun.* 24:461–468.

Edney, M. J., Rossnagel, B. G., Endo, Y., Ozawa, and Brophy, M. 2002. Pearling quality of Canadian barley varieties and their potential use as rice extenders. *J. Cereal Sci.* 36:295–305.

Evans, E. 2007. *Centrifugal impact mills*. Munson Machinery Co., Utica, NY. Published online at http://www.munsonmachinery.com.

Fast, R. B. 2001. Breakfast cereals. Pages 158–172 in: *Cereals Processing Technology*. G. Owens, ed. CRC Press, Boca Raton, FL.

FDA (U.S. Food and Drug Administration). 1999. Whole grain foods authoritative statement claim notification. Docket 99P–2209, July. FDA, Washington, DC.

FDA. 2006. Food labeling: heath claims; soluble dietary fiber from certain foods and coronary heart disease. *Fed. Reg.* 71(98): 29248–29250.

Fedec, P. 2003. Air classification. Pages 96–101 in: *Encyclopaedia of Food Science and Nutrition*, 2nd ed. B. Caaballero, L. Trugo, and P. Finglas, eds. Elsevier Science, Amstrerdam.

Fincher, G. B., and Stone, B. A. 1986. Cell walls and their components in cereal grain technology. Pages 207–295 in: *Advances in Cereal Science and Technology*, vol. 8. Y. Pomeranz, ed. American Association Cereal of Chemists, St. Paul, MN.

Fulcher, R. G., and Rooney-Duke, T. K. 2002. Whole-grain structure and organization: implications for nutritionist and processors. Pages 9–45 in: *Whole-Grain Foods in Health and Disease*. L. Marquart, J. L. Slavin, and R. G. Fulcher, eds. American Association Cereal of Chemists, St. Paul, MN.

Fulger, C. 1988. Process for producing high fiber expanded cereals. U.S. patent 4,759,942.

Gaosong, J., and Vasanthan, T. 2000. Effect of extrusion cooking on the primary structure and water solubility of β-glucan from regular and waxy barley. *Cereal Chem.* 77:396–400.

Gibson, G. 2001. *Malting*. Pages 173–203 in: *Cereal Processing Technology*. G. Owens, ed. CRC Press, Boca Raton, FL.

Goering, K. J., and Eslick, R. F. 1989. Production of beta-glucan, bran, protein, oil and maltose syrup from waxy barley. U.S. patent. 4,804,545.

Guy, R. C. E. 1991. Structure and formation in snack foods. *Extrusion Commun.* 4(Jan.–Mar.): 8–10.

Hansen, B., and Wasdovitch, B. 2005. Malt ingredients in baked goods. *Cereal Foods World* 50:18–22.

Henry, R. J. 1989. Factors influencing the rate of modification of barleys during malting. *J. Cereal Sci.* 10:51–59.

Heryford, A. G. 1987. *The influence of extrusion processing on the nutritional value of barley for weanling pigs and broiler chickens.* M.S. thesis. Montana State University, Bozeman, MT.

Huth, M., Dongowski, G., Gebhardt, E., and Flamme, W. 2000. Functional properties of dietary fiber enriched extrudates from barley. *J. Cereal Sci.* 32:115–128.

Izydorczyk, M. S., Dexter, J. E., Desjardin, R. G., Rossnagel, B. G., Lagasse, S. L., and Hatcher, D. W. 2003. Roller milling of Canadian hull-less barley: optimization of roller milling conditions and composition of mill streams. *Cereal Chem.* 80:637–644.

Jadhav, S. J., Lutz, S. E., Ghorpade, V. M., and Salunkhe, D. K. 1998. Barley: chemistry and value-added processing. *Crit. Rev. Food Sci.* 38:123–121.

Kaukovirta-Norja, A., Reinikainen, P., Olkku, J., and Laakso, S. 1997. Starch lipids of barley and malt. *Cereal Chem.* 74:733–738.

Kent, N. L., and Evers, A. D. 1994. *Kent's Technology of Cereals*, 4th ed. Elsevier Science, Oxford, UK.

King, R. D., and Dietz, H. M. 1987. Air classification of rapeseed meal. *Cereal Chem.* 64:411–413.

Kiryluk, J., Kawka, A., Gasiorowski, A., Chalcarz, A., and Aniola, J. 2000. Milling of barley to obtain β-glucan enriched products. *Nahrung* 44:238–241.

Klamczynski, A. P., and Czuchajowska, Z. 1999. Quality of flour from waxy and nonwaxy barley for production of baked products. *Cereal Chem.* 76:530–535.

Klopenstein, C. F., and Hoseney, R. C. 1987. Cholesterol-lowering effect of beta-glucan-enriched bread. *Nutr. Rep. Int.* 36:1091–1098.

Knuckles, B. E., and Chiu, M. M. 1995. β-Glucan enrichment of barley fractions by air classification and sieving. *J. Food Sci.* 60:1070–1074.

Knuckles, B. E., Chiu, M. M., and Betschart, A. A. 1992. β-Glucan-enriched fractions from laboratory-scale dry milling and sieving of barley and oats. *Cereal Chem.* 69:198–202.

Lampi, A-M., Moreau, R. A., Piironen, V., and Hicks, K. B. 2004. Pearling barley and rye to produce phytosterol-rich fractions. *Lipids* 39:783–787.

Lawrence, T. L. J. 1973. An evaluation of the micronization process for preparing cereals for the growing pig. *Anim. Prod.* 16:99–107.

Linko, Y.-Y., Mäkelä, H., and Ljnko, P. 1983. A novel process for high-maltose syrup production from barley starch. *Ann. N.Y. Acad. Sci.* 1 413:352–354.

MacLeod, A. M., Duffus, J. H., and Johnston, C. S. 1964. Development of hydrolytic enzymes in germinating grain. *J. Inst. Brew.* 70:521–528.

Marconi, E., Graziano, M., and Cubadda, R. 2000. Composition and utilization of barley pearling by-products for making functional pastas rich in dietary fiber and β-glucans. *Cereal Chem.* 77:133–139.

Masak, J., and Basarova, C. 1991. Endosperm cell wall degradation during the malting and brewing of feed barleys. *Monatsschr. Brauwis.* 44:262–267.

McGuire, C. F. 1979. Roller milling and quality evaluation of barley flour. Pages 89–93 in: *Proc. Joint Barley Utilization Seminar.* Korea Science and Engineering Foundation, Suweon, Korea.

Miller, R. C. 1990. Unit operations and equipment: IV. Extrusion and extruders. Pages 135–193 in: *Breakfast Cereals and How They Are Made.* R. B. Fast and E. F. Caldwell, eds. American Association Cereal of Chemists, St. Paul, MN.

Moreau, R. A., Flores, R. A., and Hicks, K. B. 2007. Composition of functional lipids in hulled and hulless barley in fractions obtained by scarification and in barley oil. *Cereal Chem.* 84:1–5.

Newman, C. W., Newman, R. K., and Hofer, P. J. 1994. Processing enhances the soluble fiber and hypocholesterolemic effects of waxy hulless barley. *Cereal Foods World* 39:597–598.

Newman, R. K., and Newman, C. W. 1991. Barley as a food grain. *Cereal Foods World* 36:800–805.

Niffenegger, E. V. 1964. Chemical physical characteristics of barley flour as related to its use in baked products. M.S. thesis. Montana State University, Bozeman, MT.

Norman, F. L., Hogan, J. T., and Deobald, H. J. 1965. Protein content of successive peripheral layers milled from wheat, barley, grain sorghum, and glutinous rice by tangential abrasion. *Cereal Chem.* 42:359–367.

Owens, G. 2001. Wheat, corn and coarse grain milling. Pages 27–52 in: *Cereals Processing Technology.* G. Owens, ed. Woodhead Publishing, Cambridge, UK.

Østergård, K., Björck, I., and Vainionpää, J. 1989. Effects of extrusion cooking on starch and dietary fibre in barley. *Food Chem.* 34:215–227.

Pedersen, B., Bach Knudsen, K. E., and Eggum, B. O. 1989. Nutritive value of cereal products with emphasis on the effect of milling. Pages 1–91 in: *Nutritional Value of Cereal Products, Beans, and Starches: World Review of Nutrition and Dietetics.* G. H. Bourne, ed. S. Karger, Basel, Switzerland.

Peterson, D. M. 1994. Barley tocols: effects of milling, malting and mashing. *Cereal Chem.* 71:42–44.

Pomeranz, Y., Ke, H., and Ward, A. B. 1971. Composition and utilization of milled barley products: I. Gross composition of roller-milled and air-separated fractions. *Cereal Chem.* 48:47–58.

Posner, E. S., and Hibbs, A. N. 1997. *Wheat Flour Milling.* American Association Cereal of Chemists, St. Paul, MN.

Shewry, P. R., and Darlington, H. 2003. The proteins of the mature barley grain and their role in determining malting performance. Pages 503–521 in: *Barley Science: Recent Advances from Molecular Biology to Agronomy of Yield and Quality.* G. A. Slafer, J. J. Molina-Cano, R. Savin, J. L. Araus, and I. Romagosa, eds. Haworth Press, Binghamton, NY.

Skjöldebrand, C., and Andersson, C.-G. 1987. Baking using short wave infra-red radiation. Pages 364–376 in: *Cereals in a European Context: Proc. First European Conference on Food Science and Technology.* I. D. Morton, ed. Ellis Horwood, Chichester, UK.

Sorum, D. L. 1977. A study adopting soft wheat evaluation procedures to barley. M.S. thesis. Montana State University, Bozeman, MT.

Sumner, A. K., Gebre-Egziabher, A., Tyler, R. T., and Rossnagel, B. G. 1985. Composition and properties of pearled and fines fractions from hulled and hulless barley. *Cereal Chem.* 62:112–116.

Sundberg, B., and Åman, P. 1994. Fractionation of different types of barley by roller milling and sieving. *J. Cereal Sci.* 19:179–184.

Sundberg, B., Tilly, A.-C., and Åman, P. 1995. Enrichment of mixed linked (1–3), (1–4)-β-D-glucans from a high-fibre barley-milling stream by air classification and stack-sieving. *J. Cereal Sci.* 21:205–208.

Theobald, H. E., Wishart, J. E., Martin, P. J., Buttriss, J. L., and French, J. H. 2006. The nutritional properties of flours derived from Orkney grown Bere barley (*Hordeum vulgare* L.). *Br. Nutr. Found. Nutr. Bull.* 31:8–14.

Trogh, I., Courtin, C. M., Andersson, A. A. M., Åman, P., Sørensen, J. F., and Delcour, J. A. 2004. The combined use of hull-less barley flour and xylanase as a strategy for wheat/hull-less barley flour breads with increased arabinoxylan and $(1 \rightarrow 3, 1 \rightarrow 4)$-β-D-glucan levels. *J. Cereal Sci.* 40:257–267.

Ullrich, S. E., Clancy, J. A., Eslick, R. F., and Lance, R. C. M. 1986. Beta-glucan content and viscosity of waxy barley. *J. Cereal Sci.* 4:279–285.

Vasanthan, T., and Bhatty, R. S. 1995. Starch purification after pin milling air classification of waxy, normal, and high-amylose barleys. *Cereal Chem.* 72:379–384.

Vasanthan, T., Gaosong, J., Yeung, J., and Li, J. 2002. Dietary fiber profile of barley flour as affected by extrusion cooking. *Food Chem.* 77:35–40.

Vorwerck, K. 1992. Industrial processing of barley for food. Pages 68–86 in: *Proc. Barley for Food and Malt: ICC/SFC International. Symposium.* Swedish University of Agricultural Science, Uppsala, Sweden.

Vose, J. R., and Youngs, C. G. 1978. Fractionation of barley and malted barley flours by air classification. *Cereal Chem.* 55:280–286.

Wang, L., Xue, Q., Newman, R. K., and Newman, C. W. 1993. Enrichment of tocopherols, tocotrienols, and oil in barley fractions by milling and pearling. *Cereal Chem.* 70:499–501.

Weaver, C. M., Chen, P. H., and Rynearson, S. L. 1981. Effect of milling on trace element and protein content of oats and barley. *Cereal Chem.* 58:120–124.

Wikipedia. 2007. *Hammermills.* Published online at http://www.feedmachinery.com/glossary/hammer_mill.htm.

Wingfield, J. 1980. Bread and soft wheat: recent milling progress. Pages 233–243 in: *Cereals for Food and Beverages: Recent Progress in Cereal Chemistry.* G. E. Inglett and L. Munck, eds. Academic Press, New York.

Wu, Y. V., Stringfellow, A. C., and Inglett, G. E. 1994. Protein- and β-glucan enriched fractions from high-protein, high β-glucan barleys by sieving and air classification. *Cereal Chem.* 71:220–223.

Xue, Q., Newman, R. K., and Newman, C. W. 1996. Effects of heat treatment of barley starches on in vitro digestibility and glucose responses in rats. *Cereal Chem.* 73:588–592.

Xue, Q., Wang, L., Newman, R. K., Newman, C. W., and Graham, H. 1997. Influence of the hulless, waxy starch and short-awn genes on the composition of barleys. *J. Cereal Sci.* 26:251–257.

Yeung, J. and Vasanthan, T. 2001. Pearling of hull-less barley: product composition and gel color of pearled barley flours as affected by the degree of pearling. *J. Agric. Food Chem.* 49:331–335.

Yoon, S. H., Bergland, P. T., and Fastnaught, C. E. 1995. Evaluation of selected barley cultivars and their fractions for β-glucan enrichment and viscosity. *Cereal Chem.* 72:187–190.

Zhang, H., Temelli, F., and Vansanthan, T. 2007. Phase separation and rheology of barley beta-glucan and soy protein isolate mixtures. *Cereal Foods World*. (Abstr.) 52:A72.

Zheng, G. H., Rossnagel, B. G., Tyler, R. T., and Bhatty, R. S. 2000. Distribution of β-glucan in the grain of hull-less barley. *Cereal Chem.* 77:140–144.

6 Evaluation of Food Product Quality

INTRODUCTION

Because of the physiochemical properties of β-glucan, inclusion of barley flour or milling fractions into a baked product can change the quality factors of the product. It is therefore important to measure and record quality factors, particularly when using a new or different ingredient. Methods employed by food scientists to evaluate food quality include both objective and subjective or sensory (organoleptic) procedures. Methods that can be repeated by an instrument or a standard procedure, offering an impartial record of results, are considered objective. They should be reproducible and free of individual opinion or perception. On the other hand, sensory evaluation depends on human reactions to food as perceived by the senses of sight, smell, taste, and mouthfeel. Results of objective and sensory tests of a food should correspond, if they are measuring the same component of quality. High correlation between an instrumental method and sensory testing is important in quality control. In this chapter we present a brief overview on the common procedures employed in evaluating foods that include barley as an ingredient, such as bread, other baked products, and pasta. For more detailed information, the reader is referred to Rasper and Walker (2000) and to the AACC Approved Methods (2000).

OBJECTIVE EVALUATION

Objective methods fall into the categories of chemical, physicochemical, photographic, microscopic, and physical property evaluation. Chemical methods include the analysis of nutritive value of foods either before or after cooking, as well as constituents that affect palatability, such as components that are responsible for color or flavor. Accurate methods of analysis are critical to nutritional

Barley for Food and Health: Science, Technology, and Products,
By Rosemary K. Newman and C. Walter Newman
Copyright © 2008 John Wiley & Sons, Inc.

status studies, to record information related to effects of processing and to provide reliable values on food labels (Penfield and Campbell 1990). Chemical analytical methods (moisture, protein, ash, starch, and fiber) are not described here; the reader is encouraged to refer to standards used in the cereal science field, such as those developed by the American Association of Cereal Chemists International (AACC) and the International Association of Cereal Science and Technology (ICC; www.icc.or.at). In addition, conventional routine physical testing of the raw, initial grain, such as for hectoliter weight and kernel hardness, will determine the overall potential for milling and end-use properties. Experimental milling data also provide important information on probable performance of a grain in product development on a commercial scale. Microscopic examination, especially scanning electron micrography (SEM), provides a three-dimensional image of a structure which is useful in comparing foods subjected to various treatments, such as extrusion cooking. Study of SEM photomicrographs facilitates understanding of changes that occur in components such as starch in foods subjected to various processes. Physical property evaluation such as dough testing is widely used in testing baking quality and therefore will be the focus of this chapter, particularly evaluation of bread and pasta, which are popular products to which barley flour may be added.

Evaluation of Yeast Bread

The initial consideration in testing yeast bread quality is the flour's ability to produce a "functional" dough that will produce a high-quality product. The aim of breadmaking is to convert flour (traditionally, wheat flour) into an aerated and tasty food. When moistened in dough, the gluten in wheat protein has the ability to form a cohesive mass that can stretch and trap gases from fermentation. This action is responsible for the texture and structure of bread, qualities that provide a palatable product. Achieving these qualities is a challenge when alternative grains are substituted for part of the wheat, or totally, as in the case of gluten-free breads for persons with celiac disease. Because bread is a staple food in all cultures, it is a logical product for inclusion of health-promoting ingredients, such as barley to provide soluble fiber. Barley flour alone is not appropriate for yeast breadmaking because it lacks the extensibility property of wheat gluten, therefore preventing the development of desirable crumb texture and loaf volume. Addition of barley flour or high-fiber barley milling fractions can also have detrimental effects on common parameters of quality, such as color and flavor. Therefore, experimental studies in product development commonly employ physical dough testing, functionality testing, color evaluation, and sensory testing procedures. The following is a brief description of the methods that are commonly utilized in published experimental studies on bread and other baked products using barley ingredients.

Physical Dough Testing   Physical properties of bread dough determine performance throughout the mixing process and have a pronounced effect on the finished product. Rasper and Walker (2000) divided the methodology into three phases relevant to different stages of the baking process: dough mixing, dough transformation, and gelatinization (see Figure 6.1). In the first phase, recording mixers measure and record changes in mixing resistance during the dough development time. The Brabender farinograph (C.W. Brabender Instruments Co., South Hackensack, New Jersey) is widely used, following approved Method 54-21 (AACC 2000). Two Z-shaped blades in a mixing chamber rotate at constant but different speeds and subject the dough to gentle mixing at constant temperature. An inscribed record, the farinograph curve, shows the resistance of the dough to mixing as it is formed, developed, and softened. The maximum resistance or *consistency* of the dough is adjusted by altering the amount of water added, also termed *water absorption*, which is a guide to determining the "strength" of a flour. Water absorption is defined as the amount of water required for a flour dough to reach a definite consistency at the point of optimum development. Flours from strong wheat (higher protein content) and from hard wheat require more water than do flours from weak (lower protein) or soft wheats to make a dough of standard consistency. Consistency is indicated by the maximum height of the Brabender curve, the beginning of maximum height is termed the *arrival time*, and the *stability time* is the period between arrival time and departure time when the curve begins to fall (see Figure 6.2). Another type of recording mixer,

THREE PHASE CONCEPT OF BAKING

C.W. Brabender® INSTRUMENTS, INC.	Phase I: Dough Mixing	Phase II: Fermentation & Machining	Phase III: Oven
Method	Dynamic	Static	Dynamic
INSTRUMENT	FARINOGRAPH®-E	EXTENSOGRAPH®-E	AMYLOGRAPH-E
Type of Indication	Farinogram®	Extensogram	Amylogram
Information Obtained	Absorption Mixing Time Mixing Tolerance (General Strength)	Resistance = Degree of Extensibility Maturing	Crumb Formation Gelatinization Properties Enzyme Activity
Flour Mill Correction Possibilities	Changes in Wheat Blend	Adjustment in Chemical Maturing Agents	Mill Stream Switching and/or Malt Additions (Diastating)
Bakery Correction Possibilities	Changes in Mixing & Absorption	Adjustment of Yeast Food	Malt Additions

FIGURE 6.1 Three-phase system of physical flour testing. (Courtesy of C. W. Brabender Instruments, Inc., South Hackensack, New Jersey.)

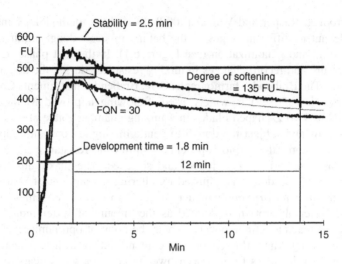

FIGURE 6.2 Representative farinogram showing some commonly measured indices. FU, farinograph units; FQN, farinograph quality number. (Courtesy of C.W. Brabender Instruments, Inc., South Hackensack, New Jersey.)

the mixograph (TMCO, Lincoln, Nebraska), originally called the Swanson mixer, provides a planetary-type action and is more suited to testing dough made from hard wheat.

The second phase of bread baking, structural transformation of dough by fermentation and maturation, employs stress–strain instruments that measure resistance of dough to extension. The extensiograph, also a Brabender instrument, records the resistance of dough to stretching and the distance a dough stretches before breaking, according to Approved Method 54-10 (AACC 2000). Another texture analyzer, the TA-XTZ (Stable Micro Systems, Surrey, England), is also effective for measuring resistance to extension. The third phase of bread baking, gelatinization, is measured by the Brabender amylograph by Approved Method 22-10 (AACC 2000). This instrument records relative viscosity, or the resistance of stirring a suspension of flour in water while temperature is increased. Starch granules swell when they approach a certain temperature, and amylose molecules seep into the liquid medium, increasing solubles concentration and therefore increasing viscosity. In its late phases, this gelatinization is also known as *pasting*. When α-amylase is present, this enzyme's action on starch results in paste thinning. Part of the starch is hydrolyzed and liquefied, with reduced viscosity, and these viscosity changes are recorded by the amylograph. Izydorczyk et al. (2001) performed dynamic oscillatory testing using a Bohlin VOR rheometer (Bohlin Reologi, Edison, New Jersey). This system provides simultaneous measurements of elastic and viscous properties of dough.

The Rapid Visco Analyzer (Newport Scientific Pt. Ltd., Warriewood, Australia) is used to measure the relative viscosity of a mixture according to Approved

Method 76-21 (AACC 2000). This procedure indicates starch heating and cooling pasting curves as well as enzyme activity, and can be used to predict the behavior of an ingredient in specific foods. Because barley can have variable starch-type differences, such as waxy or high amylopectin, there can be significant differences between varieties. Decreased amylose content in flour, such as in waxy varieties, is associated with increased hot paste viscosity of flour or starch and increased swelling of starch–water suspensions. The flour swelling volume test can be utilized for products with lower amylose, and is based on the increased swelling capacity of amylopectin (Crosbie 1991).

Functionality Testing   Experimental baking tests are commonly done in both research laboratories and industrial baking facilities. Standardized formulas and procedures such as the AACC Approved Methods can be employed to produce bread loaves for evaluation. Criteria of bread quality include loaf volume, crumb structure (grain), compressibility or firmness, and color. The comparative volume of loaves of bread can be measured easily with a rapeseed displacement test, according to Approved Method 10-05 (AACC 2000). The loaf volumeter (National Manufacturing Co., Lincoln, Nebraska) can also be employed. After the volume of a baked product has been determined, it may be divided by the weight of the product and reported as specific volume. This calculation is also facilitated if the same weight of dough is baked for each loaf. The volume index measures the area of a representative slice of bread with a planimeter or a digitizer system. The edges of the slice are traced, or a photocopy is used to determine the area. Appearance of loaves, both internal and external, provides a visual record for product comparison. Photographs or photocopies can be used as height or size comparison if a scale is included. External photographs of intact or cut loaves are also helpful in evaluating relative size, shape, and volume (Penfield and Campbell 1990).

Firmness and Structure   Firmness is often tested by compression using the Instron Universal Testing Machine (UTM) (Instron, Norwood, Massachusetts). Resistance to pressure applied to the surface of a bread slice is measured according to Approved Method 74-10 (AACC 2000). Other instruments commonly used to test bread firmness are the Baker Compressimeter (BC), using Approved Method 74-10A (AACC 2000), and the Precision Penetrometer. Baker et al. (1987) compared bread firmness measurements by the UTM, the BC, and two other instruments, and concluded that although each instrument had specific advantages, the best coefficients were between the UTM and the BC. Grain or crumb structure is important in bread quality. Consumers generally prefer small, uniform cells with thin cell walls, particularly in a white bread. Evaluation can be made visually, with a subjective comparison of photographs or photocopies. Sophisticated methods of density measurement, determining cell size distribution, such as by computerized x-ray tomography or scanning electron microscopy, are also available for evaluation of products.

Color Evaluation   Color is a huge factor in consumer acceptance of barley foods and is addressed in more detail in Chapter 7. Because there are distinct color differences in barley products due to cultivar (genetic) variation, it is essential to use objective color measurements. The Hunter color difference meter is frequently used to measure color in food products. Hunter color data are expressed in terms of L*, a*, and b*, where L* indicates brightness and a* and b* are the chromaticity coordinates. The a* value is a measure of red (+a)/green (−a), and the b* value is a measure of yellow (+b)/blue (−b) in the literature (Hunter Associates Laboratory, Inc., Reston, Virginia). Quinde et al. (2004) evaluated color brightness in barley using a spectrophotometer (CM-2002, Minolta Camera Co., Ltd., Chuo-Ku, Osaka, Japan), expressed by CIE-Lab L*. Another system of color measurement in grains is the Agtron Color Meter, which records color reflectance set on color modes of "green" or "blue" (Agtron Inc., Reno, Nevada) (Approved Method 14-30; AACC 2000).

Evaluation of Flatbreads

Flatbreads are the earliest known breads and were probably made originally from barley (Newman and Newman 2006). They have been known throughout history in the Mideast and Northern Africa as indigenous breads. Today, flatbreads take on a variety of forms, including Egyptian *balady* bread, *chapatis*, Scandinavian crisp breads, and tortillas. Each type has its own characteristics and can be evaluated accordingly, mainly for texture, compressibility, and sensory attributes. Extensibility, measured with the Instron Universal Testing machine, as well as peak load to rupture and energy level to rupture, are important in some types of flatbread (Gujral and Gaur 2005; Ames et al. 2006; Ereifej et al. 2006). In Chapter 7 we review reports of some flatbreads using barley, particularly flour tortillas, which have become popular in today's society as wraps for portable meals.

Evaluation of Quick Breads, Cakes, and Cookies

Quick breads are so called for the speed of their leavening action. Breads with chemical leavening require no fermentation time and have less gluten development than that of yeast breads. This category of baked products includes muffins, loaf breads, and biscuits. Nonwheat flours, including barley, have been used successfully as partial replacement for wheat flour in this type of product. Quick breads and muffins often contain spices, fruits, and vegetables, such as banana, berries, or pumpkin, so the alternative flours can be included without detectable effects on flavor or color. Following is a brief review of some basic quality evaluation practices for this class of baked products. For a more detailed and comprehensive guide, the reader should consult a food science text such as Penfield and Campbell (1990). In addition, the American Institute of Baking (AIB) (Manhattan, Kansas) utilizes score data sheets for these products.

Muffin quality factors are even grain, absence of tunnels, shape, and symmetry. Muffins from different formulations can be tested for volume comparison using the rapeseed displacement method cited earlier. Photocopies or photographs can be used to visually evaluate cell structure and the presence of tunnels, and also to compare shape, such as peaks or knobs on the top. Quick loaf breads are evaluated similarly to yeast breads but with more emphasis on texture and compressibility. In the United States, biscuits are rich, flaky, chemically leavened small breads, whereas the term *biscuits* in the United Kingdom refers to products known as cookies in the United States. Evaluation of American biscuits involves volume, tenderness, and texture, according to methods cited previously for breads (Penfield and Campbell 1990).

Cake quality is assessed by volume, texture, and compressibility. The consistency of cake batter can be measured as a prediction of quality in the final product, using line spread measurements or the use of devices such as a viscometer to measure viscosity or consistency. Volume measurement, the determination of volume index, is an important quality factor for cakes. Approved Method 10-91 (AACC 2000) provides templates for measuring cakes and evaluating symmetry. Appearance of the grain in cake is often used, similar to that used in evaluating the grain in yeast bread. Compression testing for crumb firmness or softness, described previously, is appropriate for cake. Approved Method 10-50D (AACC 2000) for cookie dough quality measures spread ratios, giving an indication of final size and thickness uniformity. The Bailey Shortometer (Computer Controlled Machines, Northfield, Minnesota) measures the force required to break a cookie placed across horizontal bars. Universal Testing Machine instruments may also be used to measure texture in cookies (Penfield and Campbell 1990).

Evaluation of Pasta

Durum wheat is the raw material preferred for the production of pasta. In commercial practice, a durum mill produces *semolina* flour. By definition, semolina is the fraction of milled durum remaining on a U.S. No. 100 sieve in a standard test. Flour evaluation for making pasta is generally based on kernel quality, milling performance, and end-product quality (Dick and Youngs 1988). The cooking quality of pasta is related to the quality and quantity of protein in the endosperm, and 12% protein is generally required. Any pasta product containing a partial substitution of an alternative grain such as barley is still dependent on the quality of the original wheat, particularly the gluten strength. The gluten strength of durum is measured by techniques similar to those used for bread dough: namely, use of a mixograph or farinograph. In addition, a microsedimentation test is utilized, which measures how much a ground meal swells in a sodium dodecyl sulfate and lactate solution (Dick and Youngs 1988).

In the United States, pasta products are evaluated by color, cooked weight, residue in the cooking water, firmness, and sample scoring. Color analysis usually employs the Hunter Lab color meter as well as other models that allow detection of red or brownish tones, considered undesirable. The cooked weight procedure includes cooking in distilled water, after which the pasta is drained and weighed. The cooking water is then evaporated and the dried residue is weighed, which should not exceed 7 to 8% of the dry weight. The firmness of cooked pasta is measured with an instrument such as the Instron Universal Testing Instrument (Canton, Massachusetts), following Approved Method 66-50 (AACC 2000). This procedure measures and records the work required to cut through a piece of pasta. Another evaluation of pasta is sample scoring, a statistical tabulation of at least 11 quality factors, with emphasis on protein content, color, and firmness (Dick and Youngs 1988).

In Canada (Matsuo 1988), pasta is evaluated similarly, although recent research studies have included Asian noodles with some specific evaluations for color. Images of raw noodle sheets were scanned for parameters of noodle speck size and delta-grey. *Delta-grey* defines the minimum color difference for identification of a discolored speck, and speck size is the minimum threshold size that a speck must exceed to be detected (Hatcher et al. 2005). These authors also utilized optimum cook time as a standardization for trials, and additional variations of texture measurement, maximum cutting stress, texture profile analysis, and stress relaxation tests using a texture analyzer (Texture Technologies Corp., Scarsdale, New York). Dexter et al. (2005) contributed extensive experience and knowledge to pasta quality evaluations, citing procedures of the Canadian Grain Commission (CGC, 2003). Additional instrumental techniques were reported by Sissons et al. (2005). Extensive pasta evaluation techniques have also been developed in Europe, particularly in Italy, the birthplace of spaghetti as we know it (Cubadda 1988).

SENSORY EVALUATION

Subjective or sensory food evaluation is essential in most food studies because it answers the important questions of how a food actually looks, smells, tastes, and feels in the mouth. In other words, "the proof of the pudding is in the eating" (M. de Cervantes, 1547–1616). Certainly, the inclusion of barley ingredients into a standard food product mandates the employment of sensory testing because certain barley varieties can introduce different appearance, texture, and taste, partly due to the presence of fibers such as arabinoxylans and β-glucans and partly due to phenolic compounds. Although sensory tests may appear simple and nonscientific, they are actually difficult to accomplish because they depend on human judgment, which can be affected by many factors. The Institute of Food Technology (IFT) has provided the following definition: "*Sensory evaluation* is a scientific discipline used to evoke, measure, analyze and interpret reactions to those characteristics of foods and materials as they are perceived by the senses of sight, smell, taste, touch and hearing" (IFT, Chicago, Illinois). Preparation for the tests and presentation of the products must be carefully planned and the environment standardized. We provide only a very brief description of the basics of sensory testing here. If such tests are to be undertaken, a detailed reference such as Stone and Sidel (1993) should be consulted for the theory and organization of a program. Guidelines for specific tests can be obtained from the American Society for Testing and Materials (Conshohocken, Pennsylvania). Sensory testing can be done in two general ways: groups of untrained consumers ascertain general acceptance of a product, or a trained laboratory taste panel can be used. The latter is much more involved but provides much more specific information. The consumer panel should have a large number of members, representing a cross section of a population, who are given a sample and asked to complete a very simple evaluation such as "like" or "don't like." This type of testing provides only a general idea of product acceptability; however, the results can indicate the advisability of further, more precise evaluation.

The physical area for trained taste panels should be a quiet, neutral environment with privacy booths or spaces for panelists and an arrangement for presentation of samples for testing. Training of panelists involves various aspects of taste recognition and knowledge of quality factors of the specific type of food to be tested. Standardized testing procedures should be used and results should be analyzed statistically, which requires an adequate number of panelists to assure reliable results. Three categories of tests are used in sensory evaluation: discriminative, descriptive analysis, and affective (Stone and Sidel 1993). Discriminative testing may use such simple tests as the paired comparison test, in which two samples are presented and the panelists are asked if they can discern a difference. Another is the triangle test, where three samples are presented, two of which are identical and the panelists are asked to identify the off sample. In descriptive analysis tests, a quantitative total sensory description of a product is provided, necessitating thorough screening and training of panelists. They may be asked to identify various aspects of flavor, aroma, presence of aftertaste, chewiness, crunchiness or stickiness, or any other impressions of a product. The third category, affective testing, indicates acceptance or preference of a product by making a choice between two samples presented. Another form of affective testing is the hedonic scale, in which panelists rate a product on nine points, ranging from "like extremely" to "dislike extremely." Facial image scales are also used, where panelists (sometimes children) choose a facial expression close to their reaction to a product. Additionally, the visual analogue scale (VAS) is used to measure the satiety of a food by assessing degree of hunger, fullness, nausea, desire to eat, and amount of food that can be eaten at a meal (Flint et al. 2000).

Preparation of food samples to be tested must be carried out carefully so that the appearances are identical. All samples must be equally fresh, of equal temperature, in sizes only large enough for two bites. The order of presentation is randomized and panelists should not be overwhelmed by the number of samples at a sitting. Room-temperature water should be provided for mouth rinsing between samples. Panelists are required to complete score forms to be used for analysis of results. Documentation and record keeping are essential for credibility of a sensory evaluation program. Decisions about research directions, product development, and marketing may be dependent on the results.

SUMMARY

Substitution of barley into standard wheat formulas will produce product variations due to the presence of β-glucan, polyphenols, and other unique compounds. To achieve acceptable and reproducible results, it is important to conduct and document quality evaluations of the products. A great deal of barley research has been done on yeast bread, a commonly eaten food that can provide an enhanced health benefit by inclusion of barley. Other baked goods and pasta have their own specialized objective evaluation techniques. Sensory testing is essential for consumer acceptance of new products, and requires standard procedures. Results of objective and subjective tests of a food should correspond with each other, for assurance of quality.

REFERENCES

AACC (American Association of Cereal Chemists International). 2000. *Approved Methods AACC*, 10th ed. St. Paul, MN. Published online at http://www.aaccnet.org.

Ames, N., Rhymer, C., Rossnagel, B., Therrien, M., Ryland, D., Dua, S., and Ross, K. 2006. Utilization of diverse hulless barley properties to maximize food product quality. *Cereal Foods World* 51:23–28.

Baker, A. E., Dibbon, R. A., and Ponte, J. G., 1987. Comparison of bread firmness measurement by four instruments. *Cereal Foods World* 32:486–489.

CGC (Canadian Grain Commission). 2003. *Quality of Western Wheat*. CGC, Winnipeg, MB, Canada.

Crosbie, G. B. 1991. The relationship between starch swelling properties, paste viscosity and boiled noodle quality in wheat flours. *J. Cereal Sci.* 13:145–150.

Cubadda, R. 1988. Evaluation of durum wheat, semolina and pasta in Europe. Pages 217–228 in: *Durum Wheat Chemistry and Technology*. G. Fabriani and C. Lintas, eds. American Association of Cereal Chemists, St. Paul, MN.

Dexter, J. E., Izydorczyk, M. S., Marchylo, B. A., and Schlichting, I. M. 2005. Texture and colour of pasta containing mill fractions from hull-less barley genotypes with variable content of amylose and fibre. Pages 488–493 in: *Using Cereal Science and Technology for the Benefit of Consumers: Proc. 12th International ICC Cereal and Bread Congress*. S. P. Cauvain, S. S. Salmon, and L. S. Young, eds. Woodhead Publishing, Cambridge, UK/CRC Press, Boca Raton, FL.

Dick, J. W., and Youngs, V. L. 1988. Evaluation of durum wheat, semolina and pasta in the United States, Pages 237–261 in: *Durum Wheat Chemistry and Technology*. G. Fabriani and C. Lintas, eds. American Association of Cereal Chemists, St. Paul, MN.

Ereifej, K. I., Al-Mahasneh, M. A., and Rababah, T. M. 2006. Effect of barley flour on quality of balady bread. *Int. J. Food Prop.* 9:39–49.

Flint, A., Raben, A., Blundell, J. E., and Astrup, A. 2000. Reproducibility power and validity of visual analogue scales in assessment of appetite sensations in single test-meal studies. *Int. J. Obes. Relat. Metab. Disord.* 24:38–48.

Gujral, H. S., and Gaur, S. 2005. Instrumental texture of chapati as affected by barley flour, glycerol monostearate and sodium chloride. *Int. J. Food Prop.* 8:377–385.

Hatcher, D. W., Lagassé, S., Dexter, J. E., Rossnagel, B., and Izydorczyk, M. 2005. Quality characteristics of yellow alkaline noodles enriched with hull-less barley flour. *Cereal Chem.* 82:60–69.

Izydorczyk, M. S., Hussain, A., and MacGregor, A. W. 2001. Effect of barley and barley components on rheological properties of wheat dough. *J. Cereal Sci.* 34:251–260.

Matsuo, R. R. 1988. Evaluation of durum wheat, semolina and pasta in Canada. Pages 249–261 in: *Durum Wheat Chemistry and Technology*. G. Fabriani and C. Lintas, eds. American Association of Cereal Chemists, St. Paul, MN.

Newman, C. W., and Newman, R. K. 2006. A brief history of barley foods. *Cereal Foods World* 51:4–7.

Penfield, M. P., and Campbell, A. M. 1990. *Experimental Food Science*. Academic Press, San Diego, CA.

Quinde, Z., Ullrich, S. E., and Baik, B.-K. 2004. Genotypic variation in color and discoloration potential of barley-based food products. *Cereal Chem.* 81:752–758.

Rasper, V. F., and Walker, C. E. 2000. Quality evaluation of cereals and cereal products. Pages 505–537 in: K. Kulp and J. G. Ponte, eds. *Handbook of Cereal Science and Technology* 2nd ed. Marcel Dekker, New York.

Sissons, M., Schlichting, L., Marchylo, B., Egan, N., Ames, N., Rhymer, C., and Batey, I. 2005. Pages 516–520 in: *Using Cereal Science and Technology for the Benefit of Consumers: Proc. 12th International ICC Cereal and Bread Congress*. S. P. Cauvain, S. S. Salmon, and L. S. Young, eds. Woodhead Publishing, Cambridge, UK/CRC Press, Boca Raton, FL.

Stone, H., and Sidel, J. L. 1993. *Sensory Evaluation Practices*. Academic Press, San Diego, CA.

7 Barley Food Product Research and Development

INTRODUCTION

Health statistics on chronic disease and mortality are often related by epidemiologists to food intake of a population. Official diet recommendations by government agencies and health promotion organizations define the food components needed to make changes in people's health and to change health statistics. Nutritionists and dietitians in turn promote diet changes such as the consumption of less saturated and trans fats and more dietary fiber. Consumer response to health recommendations appears to be segmented by subpopulation groups determined by lifestyle, income, emotions, tradition, and health beliefs. Food beliefs and practices are highly personal and usually not driven by accurate knowledge of nutrition or food composition. Nevertheless, trends emerge and buying patterns reflect the underlying choices by consumers. Food processors and retailers strive to meet consumer demands by developing new products with desired composition, convenience attributes, and sensory properties. Often, target markets are niches, small population segments with defined interests and groups willing to pay premiums for specified foods.

Barley has received attention from health professionals for its fiber content, particularly β-glucan, which has been shown to reduce blood cholesterol and to produce a flattened glucose response (see Chapter 8). In modern times, barley has been consumed primarily in the pearled form as a soup ingredient. To expand barley consumption to a larger portion of the diet, common food products must be considered, such as bread and pasta, that are consumed regularly and frequently. Wheat, long considered to be the "staff of life," is ideal for these foods, whereas barley by itself is not. However, the substitution of barley flour for part of the wheat flour can provide sufficient soluble dietary fiber to make a significant improvement in human health indicators. Barley is appropriate for use in a wide variety of food applications, although it poses a challenge for cereal

Barley for Food and Health: Science, Technology, and Products,
By Rosemary K. Newman and C. Walter Newman
Copyright © 2008 John Wiley & Sons, Inc.

and food scientists. Substitution of barley flour or a barley milling fraction for part of the wheat flour in yeast bread causes dilution of wheat gluten protein, which in turn causes weakening of cell structure in bread. Three obvious effects occur when barley flour is substituted for white wheat flour in breads: reduced loaf volume, darkened color, and crumb quality reduction. Color and texture changes are the most obvious effects noted in pasta products. β-Glucans, the major health-promoting components in barley, have a high affinity for moisture, adversely affecting the texture and mouthfeel of baked products. Food acceptance by consumers is strongly influenced by appearance, texture, color, and flavor, which are all characteristics that can be affected by introduction of an alternative grain such as barley. A further challenge to product developers is retention of the beneficial physiological characteristics of β-glucan throughout the processing steps so that the desired positive health benefits are maintained. This chapter is a review of current published reports of the inclusion of barley in food products. Most of the more recent studies focus on overcoming the real or perceived undesirable effects of barley in various foods. Attention is also directed toward the vast compositional variation among barley cultivars, particularly related to starch composition, as noted in Chapter 4. As development of barley cultivars continues, researchers could be directed toward those genotypes that are desirable for specific food uses, as well as modification of food-processing methods to develop superior products while maintaining health benefits of those foods.

HEALTH CLAIMS FOR BARLEY

In May 2006, the U.S. Food and Drug Administration (FDA) finalized a rule that allows foods containing barley to carry a claim that they may reduce the risk of coronary heart disease (FDA 2006). Specifically, a food made from eligible barley sources must contain at least 0.75 g of β-glucan (soluble fiber) per serving. This specification is based on eating a total of 3 g of soluble fiber daily, and included as part of a diet low in saturated fat and cholesterol and high in vegetables, fruit, and grains. Eligible barley products for the claim are whole-grain barley, pearled barley, barley bran, barley flakes, barley flour, barley grits, and whole or sieved barley meal. Table 7.1 shows the required total and soluble dietary fiber specified for these products. The barley sources are required to have been produced from dehulled or hulless clean, sound barley grain by standard dry milling processes, which may include steaming or tempering. Note that this ruling does not include barley β-glucan extracts produced through wet milling processes. The rationale for excluding wet milling–processed extracts was that the physiochemical properties of fiber and other components may be altered by the extraction processes. This possibility may be changed in the future by standardization of those processes, as discussed by Brennan and Cleary (2005). Most of the allowable dry-milled barley products specified in the FDA ruling are defined in the American Association of Cereal Chemists (AACC) glossary of barley terms, which is published in the AACC Approved Methods (AACC 2000),

TABLE 7.1   Minimum Dietary Fiber for U.S. FDA Rule on Barley Health Claims, Dry Weight Basis

Product	Total Dietary Fiber (%)	Soluble Dietary Fiber[a] (%)
Whole grain (dehulled or hulless)	10	4
Flour, grits, flakes, or meal	8	4
β-Glucan (enriched fraction or bran)	15	5.5

[a]Measurement method: AOAC Official Method 992.28 (AOAC, 2000).

and is reproduced in Appendix 3. There are two additional dry-milled products: unsifted barley meal and β-glucan fractions that are enriched in endosperm cell walls by either sifting or air classification.

BARLEY AS A FUNCTIONAL FOOD

Barley foods can be considered as functional foods, particularly since the FDA health claim approval in 2006. A *functional food* is defined as any fresh or processed food that has a health-promoting or disease-preventing property beyond the basic function of supplying nutrients. The term was introduced in Japan in the mid-1980s and refers to processed foods containing ingredients that aid specific bodily functions in addition to being nutritious (Hasler 1998). In 2005 the Institute of Food Technologists (IFT) commissioned an expert panel to review the U.S. legal standards for health-related claims and scientific standards for evaluating a proposed claim. A seven-step process for bringing functional foods to market was identified:

1. Identify a relationship between the food component and a health benefit.
2. Demonstrate the efficacy and determine the intake level necessary for the desired effect.
3. Demonstrate its safety at efficacious levels.
4. Develop a suitable food vehicle for the bioactive component.
5. Demonstrate sufficiency of scientific evidence for efficacy.
6. Communicate benefits to consumers.
7. Conduct in-market confirmation of efficacy and safety.

Barley foods are intrinsically qualified to be functional foods both for reducing cardiovascular disease risk and for modifying glycemic responses for treatment and prevention of diabetes (Lazaridou and Biliaderis 2007). Although not included in the FDA health claim, the glycemic effect may have an even more far-reaching impact on the food industry. Lifestyle changes, including dietary choices, are a daily necessity for the growing number of people with diabetes and its comorbidity metabolic syndrome.

β-GLUCAN: THE CHALLENGE OF BARLEY AS FOOD

If the primary purpose of adding barley to food products is to improve health benefits, the component in barley that is most valuable is β-glucan, either in its native form in the grain, concentrated in a milled fraction, or in an extracted purified product. Hence there is a dilemma in overcoming the characteristic physiochemical properties of β-glucan without destroying its beneficial attributes. A great amount of research has been reported on this topic in recent years, focusing on extractability, viscosity, molecular weight, and other physicochemical properties of β-glucan that affect food quality. Peter Wood of Agriculture and Agri-Food Canada has published reviews showing the relationships between solution properties and the physiological effects of β-glucan (Wood 2002; 2007). He concluded that because of the effects on β-glucan structure in processing, especially in reduced viscosity, which is crucial to physiological function, the amount of β-glucan alone cannot simply be related to the efficacy of a product. The wide range of effectiveness reported in various studies may be partially explained by the properties of β-glucan, especially viscosity. In this section we review some of the research studies reported on this subject.

Robertson et al. (1996) investigated the susceptibility of barley β-glucan polymers to disruption during cooking and digestion. Their objective was to determine the effects of various methods of β-glucan extraction and to ascertain whether or not the product extracted maintained the character of soluble fiber capable of providing beneficial physiological effects in the intestine. They determined that enzymatic treatments were more effective than either chemical or physical methods on the extractability of β-glucan from the grain. During in vitro digestion of barley, proteolytic activity was noted, suggesting the existence of a proteoglycan complex in the barley endosperm cell wall. Knuckles and Chiu (1999) treated several barley varieties with heat and chemicals to determine the effects on β-glucanase activity and molecular weights of β-glucans. β-Glucanase activities were reduced by heating, alcohol treatment, and oven heating, while hydrochloric acid and trichloracetic acid treatments reduced extractability and molecular weights of β-glucans. These results are relevant for consideration in any barley food processing venture. Sourdough bread was made with two lactobacillus starter cultures using barley flour and two milled high-fiber barley fractions (Marklinder and Johansson (1995). The amount of β-glucans in the sour doughs decreased during fermentation to different degrees (18 to 31% versus 10 to 20%) with the two starter cultures, suggesting differences in the ability of the lactobacilli to degrade the polymers. Choosing the proper lactobacillus culture is therefore necessary to maintain levels of β-glucan in the end products.

Izydorczyk et al. (2000) studied the effects of heat and physical and enzymatic treatments on total and soluble β-glucan content in flour from Canadian hulless barleys. Extractability of β-glucan from high-amylose barley was relatively low compared to normal, zero-amylose, and waxy barleys. Viscosity of barley flour slurries was affected by the proportion of soluble β-glucans, β-glucanase activity, and molecular weight of β-glucans. Hydrothermal treatments did not influence the extractability of β-glucans, but did destroy β-glucanase activity, thereby preserving the molecular weight of β-glucans and maintaining viscosity. Izydorczyk and her group studied β-glucans and arabinoxylans by producing a pearling by-product (pearlings), fiber-rich fractions, and flour from three Canadian hulless barleys with normal, high, and low amylose content. The fiber-rich fraction contained the highest amount of β-glucans, followed by flour, and then the pearling by-product. The solubility of β-glucans was inversely proportional to the levels in the fractions. The pearlings contained the highest amount of arabinoxylans. There were structural variations in both β-glucans and arabinoxylans, derived from their localization within the kernel as well as barley variety. This study exemplifies the wisdom of choosing appropriate barley genotypes and combining processing and fractionation to achieve desired products for specific applications. More recently, Irakli et al. (2004) isolated water-extractable β-glucans from six Greek barley cultivars and measured structural features and rheological properties. There were wide differences in molecular weight and corresponding differences in flow behavior, shear thinning ability and gelling capacity among samples. In addition, various sugars added to the solutions varied in their effects on gelling. The rheological behavior of β-glucan solution flow properties and gelling properties were also investigated (Vaikousi et al. 2004; Vaikousi and Biliaderis 2005). Aqueous dispersions of β-glucans underwent morphological changes during freezing and thawing. Interactions with other components were reported, as well as differences between β-glucans of varying molecular weight.

Another prolific research team, Charles Brennan and Louise Cleary of New Zealand and the United Kingdom, respectively, provided a comprehensive review of the use of β-glucan as a functional food ingredient (Brennan and Cleary 2005). This report summarized recent research and stressed the necessity for consideration of extraction procedures, molecular weight, and rheological characteristics of both extracted and native β-glucans in food. Symons and Brennan (2004a) evaluated extraction treatment on yield and composition of barley fractions, and the behavior of the fractions on starch gelatinization and pasting. Extraction with thermostable α-amylase yielded the purest β-glucan fraction. Inclusion of the fractions in starch–water dispersions significantly altered pasting characteristics related to viscosity. The effect was partially alleviated by substitution of 5% wheat starch with the fractions, which possibly reduced the enthalpy of starch gelatinization. In a complementary study, Symons and Brennan (2004b) extracted a β-glucan-rich fraction from Sunrise, a waxy hulled barley. The fraction, which contained 70% β-glucan, was incorporated into bread dough at 2.5 and 5%. Breads were evaluated for normal quality factors and also for reducing sugar release in a simulated digestion process. The rationale for measuring sugar release was to relate the soluble fiber content of the bread to a slower blood glucose response and hence a low glycemic index. The bread doughs with added β-glucan-rich fraction exhibited reduction in pasting characteristics and increased plasticity. The bread had reduced loaf volume and increased crumb firmness compared with control breads. Measurement of reducing sugars released over a period of 300 minutes was expressed in maltose equivalents as a percentage of total available carbohydrates present. A consistent decrease in sugar release was noted in the breads with the β-glucan fraction relative to the percent inclusion. The decrease in sugar release was believed to be relative to reduction in starch availability for degradation. The relationship of this phenomenon to earlier events in the dough starch hydration and pasting, as well as the presence of β-glucan-rich fraction in the bread, was unclear. Cleary et al. (2006) investigated the influence on bread baking of commercially prepared, purified β-glucan of high and low molecular weight. Yeast breads were prepared with 4.5 g of β-glucan per 100 g of bread. Breads from both treatments had stiffer dough and lower loaf volume, but the high-molecular-weight product had the stiffest dough and lowest loaf volume. Breads with both high- and low-molecular-weight β-glucan exhibited attenuated reducing sugar release during in vitro digestion. This observation is comparable to slow or "lente" carbohydrate digestion, which has significance in glycemic control (discussed in Chapter 8). The high-molecular-weight β-glucan was more susceptible than the low-molecular-weight β-glucan to destruction during bread baking, leading the authors to conclude that the low-molecular-weight product would be a more desirable bread ingredient. Electron micrographs of bread and in vitro digestion provided a comparison and understanding of the microstructure of the products.

There are numerous potential nutritional benefits of extracted barley β-glucans in food systems. Hecker et al. (1998) demonstrated the use of purified β-glucan to increase soluble fiber in tortillas. Hallfrisch et al. (2003) evaluated Nu-trimX, a barley β-glucan isolate, to lower glucose and insulin responses in healthy men and women. Glucagel, (GraceLinc Ltd., Christ Church, New Zealand), a purified β-glucan isolate from barley, forms a gelatinous precipitate that is water soluble and thermoreversible (Morgan and Ofman 1998). Brennan and Cleary (2007) prepared bread using Glucagel at the 2.5 and 5% inclusion levels. The bread with 5% isolate had increased resistance to extension compared to the control, and reduced volume. In vitro digestion of the bread indicated a decrease in reducing sugar release only from the bread with the 5% addition. The possible reasons for the lack of effect in the 2.5% product are discussed, stressing the necessity for careful consideration of the molecular and structural character of food mixtures containing starch and fiber. Glucagel has also been used in dairy products not only to add fiber but also to improve texture and microstructure. Tudorica et al. (2004) added Glucagel at several graduated levels to cheese made with milk of three different fat levels. Increased levels of the β-glucan product in milk significantly decreased coagulation time and coagulum cutting time, and appeared to form a gel network to reinforce the casein network. The most advantageous results in cheese quality occurred in the lower-fat cheeses. This same team also incorporated β-glucan at the 0.5% level into yogurt at various fat levels, with very successful results (Brennan and Tudorica 2008). β-Glucan extracted from waxy hulless barley was used successfully as a fat replacer in reduced-fat breakfast sausage (Morin et al. 2004). Another soluble fiber gel, Ricetrim, is a hydrocolloidal composite made from rice bran and barley flour (Inglett et al. 2004). Ricetrim has unique viscoelastic behavior and has been used as a fat replacer in various Thai dishes. A novel use for barley β-glucan is in edible films for food products, investigated by Tejinder (2003). The films were translucent and smooth and remained intact during immersion in water. Other barley concentrates currently on the market are Barliv, containing 70% β-glucan (Cargill, Inc., Minneapolis, Minnesota) and Cerogen, available in β-glucan content from 70 to 90% (Roxdale Foods Ltd., Auckland, New Zealand).

YEAST BREADS MADE WITH BARLEY FLOUR

The history of barley bread dates long before writing began, as noted in Chapter 1. With the advent of improved wheat flour during the nineteenth century, barley began to lose favor as a source of bread flour. In the United States, the earliest "food science" report available to these authors on inclusion of barley in bread was that of Harlan (1925), who stated that barley flour could be combined up to 20% with wheat flour without detriment to the quality of a loaf of yeast bread. Harlan stated further that up to 80% of barley flour could be incorporated into chemically leavened bread, and in both cases, barley flour darkens the color of the bread and alters the flavor. These facts are still accurate after over 80 years, and as a result, considerable research in recent years has focused on quality improvement of breads containing barley flour. Evaluating the effects of barley flour or barley components such as high-fiber fractions on the mechanical properties of bread dough is essential for predicting product success. Dissemination of this information will be useful for manufacturers seeking new raw materials for development of products with enhanced health benefits. The results of these studies provide insight into relationships between ingredient components, processing parameters, and evaluation of the end products.

Over 40 years ago, Niffeneger (1964) milled Compana barley with a laboratory Bühler mill to produce barley flour. This researcher prepared yeast bread (and other products) from 100% barley flour, 75% barley flour and 25% all-purpose wheat flour, 50% each barley and wheat flour, and 100% wheat flour. Loaf volume decreased progressively with increased proportion of barley flour. Sensory testing indicated poor texture and flavor in the breads with the higher (75 and 100%) barley levels. It was concluded that variations in the milling process, such as refined sieving methods, and a different barley variety, as well as a lower proportion of barley flour, may have produced more acceptable bread. Kim and Lee (1977) produced bread from hulless barley flour combined with hard white wheat flour in a ratio of 30% barley to 70% wheat, and included additives of 1% glyceryl monostearate and 1% calcium stearyl lactylate. The appearance, taste, and texture of the bread were very similar to those qualities of standard bread made from 100% wheat flour.

R. S. Bhatty of the University of Saskatchewan, Saskatoon, Canada was a pioneer in the milling and utilization of hulless barley in North America (Bhatty 1986). Two Canadian cultivars, Scout and Tupper, were roller-milled into bran, shorts, break flour, reduction flour, and clear flour. The break, reduction, and clear flour fractions were combined to obtain total flour yield, which was then used to determine breadmaking quality. Flour milling yields ranged from 70 to 74% at tempering moisture levels of 11 to 13%. The barley flours had a higher percentage composition of smaller particles than did the wheat flour, which may have influenced water absorption by providing a larger surface area. Whole barley meal was darker than wheat flour measured by the Hunter Lab apparatus; however, on visual observation the barley flours appeared to be similar to wheat flour in whiteness. The barleys were also pearled and ground into meal. Pearled meal of Scout barley had a higher water hydration capacity than that of Tupper pearled meal, confirming that physicochemical differences occur among barley cultivars. Both barley flours had similar pasting temperatures (64 to 66°C) to wheat flour, but lower peak viscosity as measured on a Brabender viscoamylograph instrument. Bread doughs containing 5, 10, 15, 20, 25, and 50% barley flours were prepared and tested by the mixograph instrument. Dough development times were not significantly changed by inclusion of up to 20% barley flour. However, peak height and area, and final loaf volume, were all decreased in a linear manner as barley flour was increased. This researcher concluded that 5% barley flour obtained from these hulless barleys under conditions imposed was maximum for acceptable bread loaf volume and appearance compared to standard bread made with white wheat flour. Since that time, other research groups have incorporated larger proportions of barley flour into satisfactory bread products.

A Norwegian research group (Magnus et al. 1987) made yeast loaf bread using barley of an unspecified variety which was roller-milled to 60% extraction and combined with wheat flour in proportions up to 50%. The loaf volume of the bread was reduced, and the density, measured by Instron compressibility, was increased at the higher barley flour levels of 33.3 and 50%. The addition of shortening, which was not normally used in home baking in Norway, resulted in increased volume and softness of breads. Lengthening of prefermentation time provided greater gas production, which lessened density but did not increase loaf volume. Sensory evaluation rated the flavor of bread with the lower barley level (16.7%) to be equal to that of all-wheat bread. On a hedonic scale of 1 to 5 for overall quality, taste panelists rated white wheat bread and the 16.7% barley bread with a score of 4.0, and the breads with higher barley levels 3.6 and 2.9.

The effects of sodium chloride (salt) on bread made from a 60% wheat and 40% barley flour blend were studied by Linko et al. (1984). Their rationale was that the interaction of salt ions and wheat gluten proteins reduces the water-binding capacity of gluten, and that the addition of salt to dough decreases protein solubility, among other effects. Breads were prepared by both straight-dough and sponge-dough techniques, in which a yeast–flour mixture is allowed to rise before incorporation into the dough. Loaf volumes of the wheat–barley bread increased with increased salt content, and the sponge-dough baking resulted in the best overall bread quality. Subsequently, Thual, a hulless barley, was investigated for whole-grain bread baking quality as reported by Swanson and Penfield (1988). Barley grown from two crops of this cultivar in Alaska and Tennessee was ground into whole-grain flour. Using a standard bread formula, variations of barley and salt quantities were tested for starch gelatinization and dough development properties, to select the optimum proportions. The test breads contained 20% barley flour blended with 30% whole wheat flour and 50% white bread flour. Salt was included at a level of 2.0%, since Linko et al. (1984) had reported a 20% increase in barley bread loaf volume with increased salt level. In Swanson and Penfield's study, barley decreased specific loaf volume by only 5 to 6%, and differences in texture were measured between breads made from flour of the two barley crops grown at different locations. These differences illustrate the possibility of genetic type × environmental interactions between barley crops, as noted in Chapter 4. Flavor profiles did not differ significantly from "ideal" for any of the breads. This study represented a very successful incorporation of barley flour into a standard wheat bread, which was probably due to comparison of 50% whole wheat bread with the higher density and darker color associated with breads containing barley.

Morita et al. (1996) processed two indigenous Japanese hulled barley varieties into flour by a sieving method developed in Japan. Barley flour was substituted for wheat flour at the 0, 10, 20, 30, and 40% levels in bread formulas. Barley flour was determined to possess greater water absorption capacity than wheat flour by farinograph testing, so the water content of the experimental doughs was varied relative to the barley/wheat ratios. Additionally, each type of bread was made with two different amounts of water for comparative purposes. The specific loaf volume of the breads decreased relative to the proportion of barley flour increase over the 20% level. Viscoelastic properties of dough were also affected negatively when over 20% barley was included. Sensory evaluations were conducted on the breads, using hedonic scores for gas cell distribution, flavor, taste, texture, and overall evaluation. The sensory test scores for the barley breads were highest for 10% barley flour with 90 mL of water and 20% barley flour with 170 mL of water. In some breads with higher percentages of barley flour, scores were improved with the lower water content level. Knuckles et al. (1997) produced bread in which barley flour fractions produced by roller milling and sieving and a water-extracted β-glucan fraction were included. The dry-milled and sieved fractions were substituted for 20 and 40% of wheat flour, and the water-extracted barley fraction was included at a 5% level. The dry-milled and sieved fraction contained 42% total dietary fiber and 19% β-glucan, while the water-extracted fraction contained 51% total dietary fiber and 33% β-glucan. All breads with barley fractions had increased water absorption and decreased volume. When the barley-milled fraction was included at 40%, all of the quality factors of bread were low or unacceptable. Breads containing 20% of the dry-milled fraction and 5% of the water-extracted fraction were judged by consumer panelists as acceptable for appearance, color, texture, flavor, and aroma despite of lower loaf volume and darker color. Kawka et al. (1999) added whole and fine barley flakes at up to 25% level to bread dough. The whole barley flakes at 15% reduced the loaf volume but provided the best sensory quality. In this study, the flakes added a pleasing texture and flavor to bread, and demonstrate the potential of being an ingredient in many different baked products.

A research group in India (Gujral et al. 2003) incorporated roller-milled barley flour into bread dough at 10 and 20%, together with wet gluten and ascorbic acid at two levels, in an attempt to overcome the gluten-diluting effect of barley on volume. At the 20% barley flour level, with 15% wet gluten and 20 ppm ascorbic acid, bread texture and volume were near those of control wheat bread. In addition, the additives gluten and ascorbic acid had an antistaling effect. The three ingredients in combination had a synergistic effect on bread quality factors. Dhingra and Jood (2004) roller-milled Dolma hulless barley, then blended this flour 50 : 50 with full-fat and defatted soy flour. Those 50 : 50 supplemented blends were incorporated into wheat flour at 5, 10, 15, and 20% levels for production of breads. These researchers evaluated the breads for physicochemical characteristics, nutritional composition, and organoleptic factors. The gluten content and sedimentation value of flour blends decreased and water absorption capacity increased with increases in the level of supplementary flour. All of the breads supplemented at the 20% level were nutritionally superior to control wheat bread in protein, dietary fiber, and lysine, but did not score as well by taste panels. The authors concluded that inclusion of 15% barley flour or either of the barley–soy blends produced acceptable breads with significant nutritional enhancement. Czubaszek et al. (2005) compared breads made from three spring and three winter barley cultivars grown in Poland over two years' harvests. The environment influenced protein content, dough properties, and bread quality. Pastes of winter cultivars were of higher viscosity than those of spring cultivars, and cultivar differences were measured in dough development time and speed of softening. Bread quality improved when wheat gluten, dry skim milk, and margarine were added to bread formulas.

In Australia, a new hulless barley cultivar with high-amylose starch, Himalaya 292 (later renamed BarleyMax), and two other barley cultivars were milled into flour and evaluated for breadmaking quality (Mann et al. 2005). Himalaya 292 is unique for its high levels of β-glucan (9.7%), protein (16.4%), and amylose (81.6%). The other two cultivars in this study were consistent with standard barley starch composition, about 25% amylose. Admixtures of wheat flour and barley flour were prepared in wheat–barley ratios of 80 : 20, 60 : 40, 50 : 50, 40 : 60, and 20 : 80. Himalaya 292 was used in the admixtures as whole-meal flour rather than as a roller-milled flour, due to insufficient flour yield from the material available. The effects of barley addition on the rheological properties of the doughs varied among cultivars. Himalaya 292 dough had increased strength but reduced extensibility, whereas doughs of the other two barleys had both reduced strength and extensibility. Optimum water absorption values determined by the Micro Z-arm micromixer also varied between doughs, with Himalaya 292 having a significantly higher value than those of the other barleys. Because of its unusual starch character, together with high fiber and protein contents, this barley has unique functional characteristics. Consequently, these researchers compared three different wheat flours for suitability for a wheat–barley blend to produce good-quality bread. A statistical method that uses combinations of a set of variables to derive new variables, called *principal components*, was utilized to define parameters that were the best predictors of the baking quality of wheat–barley admixtures. Two parameters that showed promise were the content of high-molecular-weight glutenin subunits of the base wheat and the extensibility of the dough measured by a texture analyzer. This study demonstrated the dynamics of wheat flour and barley flour composition and interaction in breadmaking.

The effects of substitution of cooked barley flours for part of the wheat flour in yeast bread were reported by Gill et al. (2002) in Canada. These cereal scientists used two hulless barleys, CDC Candle (waxy starch and high β-glucan) and Phoenix (normal starch and β-glucan). These were first pearled and then pin-milled into flour. The flours were stirred in water to form a slurry before cooking in a steam-jacketed kettle at 100°C for 25 minutes. The cooked and gelatinized slurry was then oven dried, reground, and sieved. Dextrose equivalents of flours were determined, to measure the reducing capacity, which influences the browning of bread crust. Cooked barley flours had higher dextrose equivalents than any of the native (raw) barley flours. Mono-, di-, and oligosaccharide composition of flours and content of these sugars was highest in cooked Candle, possibly due to breakdown of its β-glucan and starch. The cooked barley flours and their original native barley flours were made into breads, replacing wheat flour at the 5, 10, and 15% levels. All breads made with barley predictably had lower loaf volume than that of the all-wheat control. The native barley flours had decreased loaf volume with increasing barley substitution level, and at the 15% level, Candle had the lowest volume, probably due to its higher β-glucan level. The cooked flour breads at all levels of substitution had lower loaf volume than those made with native flour, except for Candle at the 15% level. The cooking of Candle flour appeared to improve its bread quality and suggests the potential of inclusion higher than the 15% level. In other evaluations of these breads, firmness increased with barley proportion increases, except for the 15% cooked Candle bread, which had firmness comparable to that of the control wheat bread. Cooked Candle flour produced dough with a soft consistency that increased at higher substitution levels. Waxy starches tend to swell and bind more water than regular starch; therefore, during baking, native Candle would probably gelatinize and bind excess water, suppressing steam generation. However, the starch in cooked Candle flour was already gelatinized, and the lack of amylose contributed to lower crumb strength, lower volume, and increased firmness. Scanning electron microscopy indicated that cooked Phoenix flour had more large starch particles than Candle. Because of Candle's high amylopectin content, particle agglomeration was minimal. The smaller particles with increased surface area may have improved starch–protein interactions and contributed to loaf volume and texture. Gaosong and Vasanthan (2000) had previously reported that Candle β-glucan undergoes molecular fragmentation when subjected to heat and shear, which could reduce viscosity and water-holding capacity. In the present study, the authors also considered the effects of starch character on firmness progression over time during storage, which is a factor of staling. The resulting breads of this study were analyzed for total (TDF), insoluble (IDF), and soluble (SDF) dietary fiber. One slice of 15% substituted Candle bread (28 g) contained 2.2 g of TDF,

31.7% more than in the control bread. The SDF content of 0.97 g per slice was 40.5% higher than the SDF of control bread.

Effects of barley and barley components on rheological properties of wheat bread dough were investigated by Izydorczyk et al. (2001). In this Canadian study, four wheat flours, varying in gluten quality from extra strong to weak, and four whole-meal barleys having varying amylose content were used in combinations to evaluate rheology properties of dough using the mixograph and dynamic rheological measurements. This experiment included incorporation of 15% of isolated and purified starch from the four barleys (normal, high-amylose, waxy, and zero-amylose) and 4% extracted β-glucan or arabinoxylan into doughs made with the four different wheat flours. The addition of either β-glucan or arabinoxylan significantly changed mixograph parameters in all wheat flours. Addition of either nonstarch polysaccharide also increased peak dough resistance, mixing stability, and work input. Both of these compounds are partially soluble, high-molecular-weight polymers, and are responsible for imparting high viscosity to aqueous solutions. They have high affinity for water and exhibit viscoelastic properties. Therefore, they may both affect water distribution in dough and may form elastic networks that contribute to dough properties (Izydorczyk and MacGregor 1998). In the present study (Izydorczyk et al. 2001), the starch additions to dough varied relative to the ratio of amylose to amylopectin, as well as in granular size. In general, the addition of starch to the various wheat flours reduced dough strength. The weakest wheat flour was affected the least by starch addition, and the stronger flours, the most. The effects of whole-barley meal addition at a 20% level to wheat also varied relative to starch type and β-glucan content, as well as the type of wheat in the combination. Normal barley reduced the peak dough resistance in all wheat flours. All of the barleys with varying amylose content improved mixing stability, and it should be noted that these barleys also had a higher β-glucan content. This suggests that the β-glucans and possibly arabinoxylans in these particular genotypes, under carefully controlled conditions, may overcome the negative effects of the dilution of wheat gluten in dough. The dynamic rheology tests in this study were an attempt to understand the role of dough components other than gluten in the viscoelastic nature of doughs. Water binding by β-glucans and arabinoxylans as well as starch type and granular size are potential factors in dough rheology and may differ by genotype. These researchers concluded that starch components in barley should be considered in addition to nonstarch polysaccharides when evaluating the effects of barley in wheat dough. In this study, the addition of starch with atypical characteristics and the nonstarch polysaccharides appeared to strengthen the wheat doughs and balance the negative effects of wheat gluten dilution by barley.

A multicenter research program in Europe, the SOLFIBREAD project, entitled "Barley β-Glucan and Wheat Arabinoxylan Soluble Fibre Technologies for Health Promoting Bread Products," had the objective of optimizing the milling process of hulless barley to produce flour for health-promoting bread products (Trogh et al. 2005). This innovative project involved milling Swedish hulless barley and extraction of arabinoxylans and β-glucans. Central to the goals of the project, arabinoxylans were separated into water-extractable and water-unextractable fractions. The water-extractable fractions form highly viscous aqueous solutions and have positive effects on dough and bread quality, whereas the water-unextractable fractions have strong water-binding capacities which are detrimental in bread-making (Courtin et al. 2001). This research group investigated endoxylanases that cleave internal linkages in the arabinoxylan backbone. Those enzymes with a higher relative activity toward water-unextractable fractions reduce water-binding capacity and increase viscosity of the aqueous phase, resulting in fragments that are beneficial in bread doughs. The functionality of endoxylanases is also influenced by endogenous endoxylanase inhibitors (Trogh et al. 2004a, 2004b), which exist in grains and can interfere with the beneficial action of endoxylanases in bread dough. Therefore, in this project, uninhibited endoxylanases were developed and used in the experiments. In one phase of the SOLFIBREAD project, wheat–barley breads were made with a composite flour of strong wheat flour and milled hulless barley flour in a 60 : 40 ratio. Xylanase that was insensitive to inhibitors was added to all doughs in graduated dosages. Flour dough and baked bread were analyzed for arabinoxylan and β-glucan content, and the molecular weight of β-glucan was also determined. The loaf volume of bread made from the composite flour without xylanase was predictably lower than that of the control wheat. The loaf volume of xylanase-supplemented wheat–barley breads was increased by 20%, and soluble arabinoxylan was also increased. Xylanase had no impact on the extractability or molecular weight of β-glucan. Therefore, the goal of producing palatable breads with increased levels of health-promoting dietary fiber components was achieved with the xylanase supplement.

Andersson et al. (2004) investigated β-glucan characteristics in dough and bread made from hulless barley milling fractions. Four Swedish nonwaxy hulless barleys were roller-milled into whole meal, flour, shorts, and bran. β-Glucans of all these samples were analyzed for molecular weight and cellotriosyl/cellotetraosyl ratios. These parameters were similar for all barleys, although some fractions had small differences. Doughs and breads were made to demonstrate the effects of flour type, water content, yeast, mixing time, and fermentation on β-glucan properties. The molecular-weight distribution of β-glucans exhibited some variations, and the average molecular weight decreased with increasing mixing and fermentation time. It was suggested that the β-glucan had been degraded by endogenous β-glucanase in both the barley and wheat flour. Baking itself did not affect molecular weight, nor did any of the processing significantly affect β-glucan structure. It was concluded by the researchers that to retain the high molecular weight of β-glucan, which is the basis of barley's cholesterol-lowering property, it is important to keep bread mixing and fermentation times short.

More recently, Jacobs et al. (2008) compared the suitability of different baking processes for incorporating high-fiber barley fractions into yeast bread. The barley fractions were obtained by pearling and roller-milling the hulless barleys CDC Candle and an experimental high-amylose line. The fiber-rich fraction contained 27.2% total dietary fiber and was added to dough at a 20% level. This was equivalent to enrichment with 4.0 g of combined β-glucan and arabinoxylan per

100 g of flour. Breads were prepared using three methods: Canadian short process, remix to bake, and sponge and dough. Two different wheats were also compared, one of which was extra-strong. The first mixing method, the short process, was considered unsuitable for making fiber-enriched bread, and the remix-to-bake method was acceptable only when the extra-strong wheat was used. The best bread product was achieved using the sponge-and-dough process because of the positive effect of sponge fermentation on gluten development and hydration. In that process, the fiber fraction was presoaked and added after the dough was fully developed. There was little difference between breads made with the two barley varieties.

FLATBREADS MADE WITH BARLEY FLOUR

Flatbreads are the world's oldest breads and are still traditional foods all over the globe (Qarooni et al. 1992). Many flatbreads are modernized versions of primitive breads, made from grain available in an area. The nomadic people were able to carry only small amounts of raw material, probably grinding grain daily to make bread. Limited equipment and scarcity of fuel necessitated a short baking period, often on a griddle or flat stove over an open fire or on hot coals. In some areas, such as the Middle East, Ethiopia, and Scandinavia, early flatbreads were made from all or part barley flour, simply because it was the most available grain. Because of the nature of flatbreads, having little or no dependence on fermentation, or volume development, they are ideal for inclusion of barley flour. Flatbreads can be made from many different grains, can be of various sizes and thicknesses, and are typically made quickly with few ingredients. An excellent resource for cereal scientists is a book by Qarooni (1996), and for the home baker, a book by Alford and Duguid (1995). In this chapter we focus on those flatbreads made with barley that have been reported in the cereal science literature.

Kawka et al. (1992) used whole barley flour ground into two levels of particle size in combination with commercial all-purpose wheat flour to make flatbread. The formula contained yeast and water in addition to the flours, and the dough was fermented for one hour before oven baking. Dough consistency was observed to be sticky in the breads made with the coarse-grained barley flour. Sensory evaluations indicated a preference for flatbreads made with fine-grained barley flour in the ratio 1 : 1 with wheat flour. Turkish flatbread (*bazlama*) was produced by Başman and Köksel (1999) using milled barley flour in combination with two different varieties of wheat flour. *Bazlama* is a flat, leavened bread about 3 cm thick, traditional in villages in Turkey and increasingly popular in urban areas. Part of the wheat flour was replaced with barley flour at the 10, 20, 30, and 40% levels. The barley flour contained 3.34% β-glucan. Addition of increasing amounts of barley flour decreased dough development time and stability. However, barley flour supplementation at the higher levels of 30 and 40% resulted in a lower farinogram mixing tolerance index and softening degree values, indicating a better tolerance to overmixing. Sensory tests indicated significantly lower scores for crumb structure and color in breads at all barley levels over 10%. All breads with barley were judged different from the control bread in mouth-feel, but were also judged equal in flavor, aroma, and appearance, and all were deemed acceptable. Hunter Lab color values indicated an increase in grayish tints as barley percent increased. Both β-glucan and dietary fiber increased in relation to the level of barley inclusion in the breads. The authors concluded that the inclusion of barley provided a potential for improvement in the health-promoting composition of *bazlama*.

A different type of Turkish flatbread, *yufka*, was investigated by the same group (Başman and Köksel 2001). *Yufka* is an unleavened flatbread consumed in Middle Eastern countries. The dough for this bread needs to be flexible, so it may be rolled. The experimental design of this study was very similar to the previous *bazlama* study, and the same wheat and barley cultivars were used. The bread formula differed by the absence of yeast, and the procedure differed in that the dough was rolled thinner for *yufka*. As with *bazlama*, the *yufka* breads decreased in sensory scores relative to the increasing proportion of barley flour, but all breads were considered acceptable by the panelists. β-Glucan and dietary fiber levels increased in proportion to the levels of supplemental barley flour.

Chapatis, unleavened flatbreads originating in India, are traditionally made from whole wheat flour, giving a nutritional advantage over some other flatbreads. They are often baked on hot flat iron plates or griddles rather than in the oven. Anjum et al. (1991) evaluated six high-lysine barley lines for nutritive value and functionality in preparing *chapatis* in Pakistan. Proportions of barley flour from 5 to 100% of the wheat flour were incorporated into *chapati* doughs. At the 10% level of substitution, farinograph characteristics were comparable to all-wheat controls, but higher levels of barley produced inferior dough structure and lower-volume *chapatis*. An Indian group (Sood et al. 1992) investigated the use of a hulless barley, Dolma, in *chapatis*. The barley grain and a control wheat were stone milled and sieved to obtain 70% extraction flour. *Chapatis* were prepared with proportions of barley flour from 10 to 50% of the wheat flour. Water absorption capacity increased relative to barley meal content, as predicted. The color and appearance of *chapatis* were evaluated as good up to the 30% level of barley. Flavor and texture of *chapatis* were rated acceptable up to the 40% substitution of barley flour.

Gujral and Pathak (2002) prepared *chapatis* from various composite flours, including barley, replacing the base whole wheat flour by 10, 20, 30, 40 and 50%. The objective of this study was to compare the effects of alternative flours on extensibility using the Instron Universal Testing Machine. During storage, *chapatis* typically become stale and tough, so the study included testing breads that were both freshly baked and some stored for 24 hours. Control wheat *chapatis* were soft and extensible (8.117 mm), but after 24 hours the extensibility decreased to 2.213 mm. Maximum force required to rupture the wheat *chapatis* increased after storage, and the energy required to cause rupture decreased. The addition of barley flour increased the extensibility of *chapatis* to 12.91 mm at the 20% level, but at higher levels of barley the extensibility began to decrease.

After storage, however, the extensibility remained significantly higher than the control. *Chapatis* containing barley flour also required higher peak force to rupture and greater energy to rupture, factors that remained higher after storage. The authors suggested that the β-glucan in barley with high water absorption properties was responsible for the preservation of starch character, thus maintaining extensibility during storage. This research group later evaluated *chapatis* made with barley flour incorporated into wheat dough at the 10 and 20% levels (Gujral and Gaur 2005). In this study, glycerol monostearate, believed to decrease starch recrystallization and salt, believed to retard staling, were added to the *chapati* dough. Breads were measured for extensibility, peak load to rupture, and energy to rupture, both fresh and after storage, as in the earlier study. Barley flour at the 20% level increased extensibility in both fresh and stored *chapatis* without the presence of the other additives. The combination of 10 to 20% barley flour with 0.25 to 0.5% glycerol monostearate and 0.5 to 1% NaCl improved *chapati* extensibility further, especially in the stored state, thus restraining staling.

Another type of flatbread, *balady*, was investigated in Jordan (Ereifej et al. 2006). *Balady* bread, sometimes called *pita* or *pocket bread*, is commonly eaten in Egypt and the eastern Mediterranean area. The objective of this study was to consider the possibility of replacing part of the wheat, which tends to have low crop yield in the region, with barley, which grows better in less fertile land. Two barley varieties, Rum and ACSAD, were ground and sieved and then combined with two wheat flours at levels ranging from 0 to 100%. The two wheat flours differed in protein and carbohydrate content, as did the two barley whole grains and sieved flour. Hence, the combinations of grains with wide variations in components could be expected to provide variable results. The breads were prepared according to standard formulas and then measured for loaf volume, thickness, and pocket formation. As predicted by the researchers, breads made with 15 or 30% of either barley flour were considered acceptable for those parameters. Sensory evaluations included properties that determine acceptance of bread by consumers, including flavor, odor, loaf color, homogeneity of color, uniformity of loaf, chewiness, absorption of water, and loaf peripheries. Differences were found between the two barley varieties as to the maximum amount of barley flour that could be combined with wheat. There were also differences between the bread quality from the two wheat varieties. This interaction effect exemplifies varietal differences between grain crops to be used for breadmaking, which is well recognized with wheat. Another study on pita bread made with barley was reported by a different research group in the United Arab Emirate (Ragaee and Abdel-Aal 2006). Whole-grain barley meal was blended with wheat flour in different proportions. Rapid Visco Analyzer tests were used to help formulate flour blends with desirable pasting properties. The combination of 15% barley meal with 85% wheat flour proved acceptable for satisfactory pita bread in this study.

Tortillas, a common flatbread in North America, particularly in Mexico and the southwestern United States, were originally made with maize (corn) meal, but wheat flour is also now utilized. Efforts have been made to fortify or use alternative grains in tortillas to increase dietary fiber. Wheat tortillas from both white and whole-grain flours have become very trendy in modern diets as "wraps" for sandwich fillings. Tortillas are an excellent bread type for supplementation with health-promoting ingredients, as there is little dependence on volume, height, or cell structure, and the major quality factors to strive for are tenderness and rollability. Suhendro et al. (1993) added milk, oilseed proteins, and gluten to corn tortillas. Others have supplemented the corn with sorghum or triticale (Torres et al. 1993; Serna-Saldivar 2004). Hecker et al. (1998) investigated the use of an extract of β-glucan from waxy hulless barley that contained 56% TDF as a fiber supplement in flour tortillas. β-Glucan extract was incorporated into tortilla dough to provide 2 g of soluble fiber in each serving. The supplemented tortillas were slightly more tender than all-wheat controls, and rollability scores were similar to controls over a five-day period. Ames et al. (2006) prepared flour tortillas with 100% barley flour. Several barley cultivars were evaluated before selection of an appropriate genotype. The criteria used were the amount of water required for optimum dough consistency and avoidance of stickiness during sheeting. Barley tortillas were evaluated by a sensory panel as well as objective mechanical measurements with a texture analyzer. Significant correlations were found between sensory and instrumental methods. Consumers preferred tortillas with lower break ability, lower dryness, higher rollability, and greater ease of compression. These qualities were deemed predictable by using instrumental techniques. The waxy hulless barley cultivar CDC Candle, which has a low amylose and high β-glucan content, was made into superior tortillas, as well as yeast bread, as mentioned previously. Guo et al. (2003) reported that low-amylose (waxy) wheat flour produced tortilla dough with enhanced extensibility. Waxy starch appears to be an asset in tortilla structural qualities.

CHEMICALLY LEAVENED BAKED PRODUCTS MADE WITH BARLEY FLOUR

Barley has been incorporated successfully into chemically leavened baked products, even as 100% of the flour content. However, as with the studies above, genotype of the barley plays a large part in certain quality factors. Klamczynski and Czuchajowska (1999) compared waxy hulless and nonwaxy hulled barleys for baking quality in quick breads and other products, and reported cultivar differences in volume and crumb texture. Newman et al. (1990) prepared muffins from three isotype hulless barleys, two of which had waxy starch, derived from Compana, a hulled barley. The barleys, including the parent Compana, were roller-milled into flour and incorporated into a standard plain muffin formula using 100% barley. Wheat muffins were prepared as a control. Agtron color reflectance values indicated that all of the barley flours and muffins were darker than the wheat control. Three of the barley flours reached peak viscosity with the Rapid Visco Analyzer at a lower temperature than wheat, indicating the waxy starch influence. The wheat muffins appeared to have the largest volume, although none of the barley muffins were statistically different from the wheat.

Taste panelists were unable to distinguish among muffins made from the three hulless and/or waxy cultivars. Panelists preferred barley to the wheat muffins and commented on moistness, sweetness, and "bran" taste. This study demonstrated that although there were wide differences in starch and fiber composition among the barleys, acceptable results can be achieved in a chemically leavened product.

Hudson et al. (1992) studied muffins made with a high-fiber barley fraction. The barley, a covered normal-starch type, was dehulled, then milled and sieved to yield coarse particles high in β-glucan (18.9%). This flour was incorporated into a muffin formula in a 40 : 60 ratio with all-purpose wheat flour. This recipe used enhancement ingredients, including honey, orange extract and peel, cinnamon, and chopped dates. The 40% inclusion level for the high-fiber barley fraction was considered maximum for an acceptable product. The muffin volume was adequate, but muffins lacked the characteristic cracks seen on the tops of commercial muffins. Taste panelists rated the barley muffins at 6.4 on a scale of 1 to 9. Soluble dietary fiber in the muffins was 3.49% and β-glucan was 3.79%, both substantially higher than in a commercial oat bran muffin mix. Berglund et al. (1992) tested a variety of products, including muffins, biscuits, cookies, and granola bars, using Wanubet, a waxy starch hulless cultivar ground into whole-grain flour and also processed into flakes for some of the products. This broad-spectrum study evaluated products with barley flour inclusion from 26 to 100%. Objective evaluations included volume, specific volume, color using the L*, a*, b* scale, and cookie spread. Typically, the volume of barley products was slightly lower that of wheat controls, and the color was darker. In sensory evaluations, all barley products compared favorably to controls, and granola bars made with barley flakes were preferred by panelists to those with rolled oats. The authors predicted a high rate of success for barley utilization in high-fiber food products.

Newman et al. (1998) selected the shorts fraction of roller-milled Shonkin, a waxy hulless barley in Montana, to produce yeast bread, biscuits, cookies, and muffins. The shorts fraction, which contained 18.2% TDF and 8.8% SDF, was incorporated into each product at a level calculated to double the SDF. Each product was prepared both with and without the fraction, and those without the fraction contained a 50 : 50 blend of white and whole wheat flours, in order to provide a similar visual and texture comparison with the fiber-enriched products. All products were evaluated objectively for volume and dimension, and by sensory tests. The only fiber-enriched product with reduced volume and dimensions was the yeast bread. In sensory evaluations, total scores between fiber and standard products did not differ except for muffins, which were described as "moist." Hedonic scores for overall quality were similar for all products. SDF in barley-enriched products was approximately double that of the respective standard items.

Other baked products, such as cookies and snack bars, can be made successfully with all or partial barley flour replacing wheat flour. Sugar snap cookies were prepared using 100% barley flour that had been milled on a Bühler experimental mill (McGuire 1984). Soft wheat methodology was used for the cookies, with several different barley cultivars. Cookie diameters varied widely with different cultivars, as did alkaline water-retention capacities, which were typically high. Cookie spread was correlated with flour ash content, possibly a function of milling yield and degree of separation of bran fractions. Protein content did not affect the spread factor of the cookies. Klamacznski and Czuchajowska (1999) compared flours from waxy and nonwaxy barley for the production of sugar cookies. The two types of barley differed as to milling fractions, break, and reduction. Cookies made from nonwaxy barley and break flour were larger in diameter than those made from waxy barley and other flours. It was believed that amylose content, particle size distribution, and ash and β-glucan level were factors influencing cookie quality. Ragaee and Abdel-Aal (2006) replaced wheat flour with 30% whole-grain barley meal in cake and cookies. They reported minimal differences in volume measurements or sensory evaluations between the barley products and wheat controls. Botero-Omary et al. (2004) prepared chocolate chip cookies containing 50% flour milled from Sustagrain, a high-β-glucan (15%) barley flour. Cookies were evaluated for appearance, color, flavor, texture, and overall acceptability using a nine-point hedonic scale. Panelist responses were separated into groups, and the most favorable ratings were seen among males, Hispanics, and people who were conscious of their fiber consumption.

PASTA MADE WITH BARLEY FLOUR

Pasta is known by many different names, with spaghetti, noodles, and macaroni being the most commonly known. Modern popular menus and recipes, however, refer to fettucini, penne, rigatoni, and ziti as well as many more shapes and styles. The popularity of pasta has grown enormously in recent years. The current interest in whole grains and fiber-rich foods has sparked interest in development of new pasta products containing whole wheat and supplemental fiber sources, including barley flour or barley milling fractions.

Addition of high-β-glucan fractions from barley to pasta was undertaken by Knuckles et al. (1999) at the USDA Western Regional Research Center. Two barley fractions were incorporated into pasta dough, which was dried, cooked, and evaluated for quality factors. The cooked pastas contained 10 to 20% TDF and 4 to 8.6% β-glucan. The barley pastas all had acceptable sensory quality, although they were darker in color than the all-wheat control. These researchers concluded that pasta could be enriched with concentrated β-glucan fractions to enhance soluble fiber content and achieve acceptable products.

An Italian group (Marconi et al. 2000) prepared functional spaghetti with high-fiber fractions from barley. Two commercial barleys were pearled, and the resulting pearlings were milled and sieved to prepare the high-fiber fractions. Composite flours were made using durum wheat semolina blended with two fractions, each substituted for 50% of the total, with 5% vital wheat gluten added. One additional composite flour was made with milled residual pearled barley kernels. Spaghetti was produced and dried using standard methods. The color values of the barley pastas were darker by L*, a*, and b* measurements than the semolina pasta control. The cooking quality of all pastas with barley fractions was acceptable, and that made from milled, pearled barley kernels was almost identical in quality factors to the all-semolina pasta. The objective of this study, to produce functional pasta rich in dietary fiber, was met. The barley pastas contained 13 to 16% TDF compared to 4.0% in the control, and 5.9 to 7.7% SDF compared to 2.7% in the control. Barley pastas contained 4.3% β-glucan. These values met the FDA requirements of 5 g of TDF and 0.75 g of β-glucan per serving (56 g in the United States and 80 g in Italy) to allow these pastas to be labeled a "good source of dietary fiber" as a measure to reduce the risk of heart disease.

Dexter et al. (2005) reported the use of hulless barley milling fractions in spaghetti. This study was unique in that barleys with variable levels of amylose and β-glucan were utilized. The barleys used were Falcon (normal starch), CDC Alamo (zero amylose), and a high-amylose experimental line. The barleys were pearled and roller-milled into flour and fiber-rich fractions. Spaghetti was produced with semolina wheat flour enriched with 20% and 40% barley flour and separately with the fiber-rich fractions sufficient to increase β-glucan in the spaghetti by 2%. Predictably, the color of the 40% barley pasta was darker, but pearling before milling produced a lighter-colored product. The higher level of barley flour made spaghetti less firm, particularly in the case of that made with zero-amylose barley. Generally, the quality of the fiber-enriched spaghetti was very acceptable. Other researchers (Cleary and Brennan 2006) incorporated a high-β-glucan fraction derived from barley into pasta at 2.5, 5.0, 7.5, and 10% of wheat flour. Solid loss was greater during cooking, pasta swelling increased, and loss of hardness occurred compared to the control pasta. There were two innovative aspects of this study. The β-glucan fraction attenuated reducing sugar release during in vitro digestion relative to the level of fiber fraction added. This was believed to be due to the water-binding capacity of the β-glucan. This feature is of great interest for achieving a low-glycemic-index value for a product. Foods with low glycemic responses are desirable for people with diabetes or impaired glucose tolerance. In addition, examination of the pasta microstructure and characterization of starch gelatinization suggested that the fiber fraction affected the starch–protein matrix of the pasta.

Noodles are very popular among consumers of Asian ancestry. Fifty percent of wheat consumed in Asia is in the form of noodle products (Hoseney 1992). The two general types of noodles are classified as white (Japanese type) or yellow alkaline (Chinese type). In Korea, Change and Lee (1974) prepared noodles from wheat with added barley flour. Barley at the 20 or 30% level of inclusion resulted in acceptable noodles except for color, which was considered dark and grayish. Kim et al. (1973) used hulless barley flour for noodles, which were of good quality. Han (1996) incorporated flours from five Montana barley varieties with hard wheat to make Chinese noodles. The barleys varied in β-glucan content from 3.57 to 6.23%, and barley flour was added at a level of 5% of the wheat flour. Color changes in cooked noodles were minimal at this low level of barley.

An interesting aspect was the color changes in dough sheets during the first hour of holding, possible due to water distribution in the dough. After 7 hours the dough color stabilized. This researcher used two different wheat varieties and observed texture differences. Hardness in cooked noodles was increased when higher-protein wheat was used.

Udon, Japanese-type noodles, were investigated by Baik and Czuchajowska (1997). This comprehensive study was done with hulless waxy and nonwaxy barleys abraded in a pearler to remove 10 or 20% of the kernel. Barley kernels were then ground to produce barley flour, which was blended with wheat in the ratio of 15% barley and 85% wheat. All flours were evaluated for pasting characteristics, which are considered important for udon noodles. The barleys that were abraded to 20% had higher amylograph peak viscosities due to increased starch content, and this was most evident in the waxy varieties. The wheat–barley blended flours all exhibited increased amylograph breakdown. Noodles were evaluated by texture profile analyses for hardness, chewiness, and other parameters. Texture profile analysis parameters were lowered in noodles made with waxy barley, especially that which had been abraded to 20%. A shorter cook time was required for all of the noodles containing barley. Scanning electron microscopy revealed more open internal structures of noodles with barley. The authors attributed this to starch gelatinization and water retention, causing an open network to develop from the outside of the noodle surface. Barley flour-containing noodles were darker in color, but to a lesser degree than those made with the 20% abraded barley.

Yellow alkaline noodles enriched with eight hulless barleys of varying starch types were reported by a Canadian group (Hatcher et al. 2005). Barleys included were normal starch, waxy, zero-amylose, waxy, and high-amylose types. A spring wheat variety, Canada Prairie, was used. All barleys and the wheat were pearled and roller-milled into flour. Barley flours were added to wheat flour at a 20% level, and the zero-amylose barley was added at a 40% level. Noodles were evaluated for color by Hunter Lab values and appearance related to visible specks by scanned images analyzed for delta-gray and speck size (Hatcher et al. 1999). All raw noodles containing barley flour had significantly reduced brightness (L*) and yellowness (b*), elevated redness (a*), and increased number of specks per area compared with all-wheat-flour noodles. The number of specks also increased to a greater degree with the higher level of barley flour. Color intensity increased in the cooked barley noodles. Noodle firmness was unaffected by barley at the 20% addition, but did increase at the 40% level. Chewiness decreased in waxy and zero-amylose barleys and increased in normal and high-amylose cultivars. These researchers commented that the effect on the quality of yellow alkaline noodles is complex when barley flour is added, particularly in relation to the starch composition. Waxy and zero-amylose starch decreased chewiness and relaxation-time parameters, while normal and high-amylose starch increased firmness and chewiness. Also, the presence of β-glucan as well as starch type could contribute to the texture and viscosity of noodles.

Izydorczyk et al. (2005) selected a waxy and a high-amylose barley from the study above (Hatcher et al. 2005) to enrich both white salted and yellow alkaline

Asian noodles. These barleys were roller-milled to produce fiber-rich fractions with 22% β-glucans and average 34% TDF. The fractions consisted mostly of irregularly shaped fragments of endosperm cell walls, which were porous in structure. Noodle dough was prepared with 25% fiber-rich fraction and 75% wheat flour. Inclusion of fiber-rich fractions increased all viscosity parameters of the noodle dough, due partially to water-holding properties. Additional water was required in doughs to attain proper consistency. All noodles with fiber-rich fractions were darker than control wheat noodles and contained more brown specks. Cooking time for both white and yellow noodles with fiber-rich fractions was reduced by 50%. Cooked yellow noodles containing fiber-rich fractions had good firmness and resistance to compression, whereas the white noodles with fiber-rich fractions were less firm and less chewy. In vitro digestibility of barley noodle starch indicated decreased glucose release, which suggests a low glycemic index value. This same research group (Lagassé et al. 2006) investigated the Canadian hulless barleys reported by Hatcher et al. (2005): genotypes with normal, waxy, zero-amylose, and high-amylose starch. In the Lagassé study, barleys were milled into flour and blended at 20 and 40% levels with wheat flour for preparation of yellow alkaline noodles. All barley noodles processed normally, although work input varied somewhat relative to barley starch type. As predicted, color was affected negatively. The polyphenolic activity of the barley flours was measured as a possible answer to color problems. All the barleys possessed higher phenolic activity than that of the wheat flour. Barley may contain 1.2 to 1.5% polyphenols, which cause discoloration in baked products (Quinde et al. 2004). Texture of barley noodles was strongly affected by starch type. Noodles containing high-amylose or normal starch had bite and chewiness equal to wheat noodles, while waxy and zero-amylose types had lower texture quality. These researchers concluded that white noodles made with 40% barley flour contained over 2.0% β-glucan and offer a healthful choice to consumers. The color factor represents another anomaly between health benefits and consumer preferences. The presence of polyphenols provides additional health benefits, such as antioxidant activity and prevention of cellular defects. Although there is a preference for white color in noodles, health-conscious consumers may eventually accept a darker color.

Although not a pasta, Chinese steamed bread is a popular food in Asia and among immigrants in many parts of the world. As with other wheat-based products, it contains less than desirable amounts of fiber. Newman et al. (1994) prepared typical Chinese steam bread, substituting 10, 15, and 20% of the wheat flour with barley flour. The flours were obtained by roller milling a hard white winter wheat (MT 7811) and a waxy hulless spring barley (Merlin). The wheat and barley flours contained 2.4 and 7.7% TDF and 13.2 and 10.3% protein, respectively. When barley flour was substituted for wheat flour, farinograph parameters were changed and the dough generally became weak, dough development time became shorter, mixing tolerance index values increased, and dough stability time became shorter. Water absorption time increased slightly with the substitution of barley flour for wheat flour. Crumb texture and color decreased, but in a linear fashion with increased increments of barley flour. Volume decrease was a more visible effect of barley flour substitution. Volume expressed as milliliters per 100 g of dough were 298, 292, 288, and 278 for all wheat flour, and 10, 15, and 20% barley flour, respectively. However, a trained taste panel of eight Chinese persons determined that there was no difference in eating quality among the four breads.

BARLEY BREWERS' AND DISTILLERS' GRAINS USED IN FOODS

Spent brewers' grain, a by-product of the malt industry, has historically been utilized as livestock feed. The spent grain consists mainly of the pericarp and hull portions of barley and the nonstarch parts of adjunct grain such as corn or rice. Although "spent" in terms of fermentable starch, the mixture is concentrated in protein and TDF and is suitable for use as a supplement in food products. The major difference between spent grain and barley flour or milled fractions is a lack of starch and SDF. Prentice and D'Appolonia (1977) used commercial dried spent brewers' grain as a fiber supplement in bread made from 70% white and 30% whole wheat flours. Spent grain was given heat treatments at three temperatures and incorporated into bread at 5, 10, and 15% of flour. Consumer panels favorably accepted breads made with spent grain at 5 and 10% levels if the spent grain was preheated at no more than 45°C. Dreese and Hoseney (1982) further separated spent grain into hulls and bran fractions. The bran fraction was added to wheat flour dough at 15%. Results indicated mixing problems and reduced volume in the breads. An enhanced process of drying and milling spent barley brewers' grain resulted in a new product, barley bran flour, containing 70% TDF and 18.5% protein. This product was used to replace 15% of the flour in wheat bread (Chaudhary and Weber 1990). The breads contained 8.9% TDF and were very acceptable compared to breads made with other fiber supplements. A barley by-product from ethanol production, distillers' dried barley grain has been investigated as a fiber supplement (Eidet et al. 1984). Quick breads containing graduated levels of a flour fraction of distillers' grain were prepared and evaluated. Volume index of breads varied only at the 10 and 15% levels of distillers' grain content. Taste panelists judged breads to be acceptable in flavor and texture at both the 5 and 10% levels, and color in the pumpkin and carrot breads was less adversely affected because of their intrinsic color. Dawson et al. (1985) produced muffins containing 15% barley distillers' grain which had been defatted, resulting in flavor improvement of the products. Distillers' grain flour was also incorporated into sausage as a fiber supplement (Levine and Newman 1986). Mildly seasoned Polish sausage was prepared with 3.5% of whole and fractionated distillers' grain and compared with a product containing soy isolate as a control. Water solubility and emulsifying capacity were poor in the distillers' grain sausages, although acceptability by a taste panel was not different from that of controls with soy isolate.

PEARLED AND OTHER FORMS OF BARLEY

In Japan, Korea, and other Asian countries, barley has been used as a rice extender for many years, especially when rice was in short supply. When rice was plentiful and cheaper, there was less demand for barley. In recent years, however, the health implications for barley have renewed interest in barley as a rice extender (Ikegami et al. 1996). In Korea, the word *bob* means cooked rice, and a mixture of barley and rice is called *barley bob*. It is usually prepared by partially precooking pearled barley to be added to rice being cooked (B.-K. Baik, personal communication). In some cases the barley grains are split in half to equalize cooking times (Yang et al. 1999). It is the personal experience of the authors of this book that brown rice and pearled barley require about the same cooking time and can be cooked together. The same is true for white rice and Quick Barley, a product available in North America (The Quaker Oats Company, Chicago, Illinois). Thus, an Americanized version of barley bob provides enhanced soluble fiber for any use of rice in meals.

Edney et al. (2002) investigated the pearling quality of 25 Canadian barleys for potential use as rice extenders to meet market demands from Asia. The quality factors evaluated included pearling times, color of pearled grains, steely kernel counts, and crease width. In general, waxy genotypes produced higher-quality end products. The L* brightness value was a good predictor for kernel whiteness as well as endosperm texture, and the particle size index as measured by the single kernel characterization system was an indicator of kernel texture. Klamczynski et al. (1998) examined two nonwaxy hulled and two waxy hulless barleys that were abraded at different levels. Barley kernels were evaluated for properties that relate to quality for use as a rice extender. Nonwaxy barleys differed in degree of seed coat removal by abrasion; with resulting differences in maximum water imbibition due to soaking over time. The extent of starch gelatinization during cooking was measured. Barleys abraded to a greater degree allowed water and heat to penetrate more quickly during cooking. The results of this study illustrated that abrasion degree affects the cooking quality of kernels.

Bulgur, a semi-whole-grain product traditional in the Middle and Near Eastern countries, can be produced from barley as well as from wheat (Köksel et al. 1999). Bulgur can be cooked and served like rice, combined with meat as a main dish, or made into a cold salad called *tabbouleh*. Three Turkish barley cultivars were soaked, cooked, and dried. Grains were then dehulled, cracked, and sieved to remove fine particles. The resulting bulgur was analyzed for nutritive value and found to have increased β-glucan (about 5%) and lowered levels of phytate phosphorous, suggesting better bioavailability of certain minerals such as zinc.

Tarhana is a traditional food from the Middle East and Eastern Europe made from whole-meal ground wheat or barley and then prepared into a sort of dried soup mix. In Turkey, flour is mixed with yogurt, yeast, chopped vegetables, and spices. The mixture is allowed to ferment for up to seven days, then spread into sheets and dried, traditionally in the sun. After drying, *tarhana* keeps very well as a staple food product. Erkan et al. (2006) prepared *tarhana* with three barley cultivars, two hulled and one hulless, which were roller-milled into flour. The barley flours were combined with yogurt, yeast, onions, green and red peppers, spices, and salt. The mixture was fermented, covered, at 30°C for five days. Batches were also prepared with wheat and a 1 : 1 wheat–barley blend. After fermentation, *tarhana* was dried at room temperature before grinding and sieving. The β-glucan content of the barley flours ranged from 2.8 to 4.25% and in the dry barley *tarhana* from 2.4 to 3.6%. Soups were made by mixing 40 g of *tarhana* with 500 mL of water and simmering for 10 minutes with constant stirring. The differences between products were not substantial enough to affect sensory evaluation. The use of barley flours affected color values, with some viscosity differences in the soups. The authors addressed the possibility of some destruction of β-glucan during fermentation, and concluded that despite small losses, barley flours were still a desirable ingredient to produce *tarhana* with more healthy properties. *Kavut* is another indigenous cereal product consumed in the eastern regions of Turkey. Whole-meal wheat and barley grains are generally used in the production of this sweet food. The whole grains are first roasted and then ground by stone milling. Milk, sugar, and butter are kneaded into the coarse whole grain to form balls or patties. Karaodlu and Kotancilar (2006) made *kavut* with Turkish wheat and barley that were roasted using three different heating periods at 250°C. Grains were ground and used in different proportions, combined with the other ingredients, then formed into shapes, wrapped, and stored. The length of the grain roasting process affected the softness and color of *kavut*, but not to the extent of altering taste panel acceptance. However, increased proportion of barley in *kavut* generally decreased the overall sensory quality. The best level of taste and flavor of wheat–barley blends was 1 : 1, but panelists still preferred all-whole wheat *kavut*.

The word *porridge* in modern times means a soft food made of cereal or meal boiled in water or milk until thick. The original term *pottage* referred to a thick soup or stew made from vegetables and/or meat (*Webster's New World Dictionary*, 1986 edition). The latter meaning can therefore include many traditional cereal based foods that in earlier times could have served as one-dish meals. Today, *porridge* generally refers to a hot breakfast cereal that is served with milk and sometimes with a sweetener. A thinner form, *gruel*, is used as a weaning food in many cultures. Pedersen et al. (1989) investigated the use of a Danish high-lysine barley cultivar to enhance the nutritional value of barley gruel for infants in developing countries. The high-lysine barley contained 4.9 g of lysine per 16 g of nitrogen, compared to 3.8 g in the control normal barley. This high-lysine barley can provide a safe level of protein intake for infants and small children based on World Health Organization recommendations. The rationale was that if 60% of a 1-year-old child's energy intake is derived from barley gruel, the gruel will provide 94% of the child's total protein requirement, twice that provided by a normal barley or wheat, and three to four times that of maize. To meet 60% of energy requirements, a child would need to consume at least four meals of 200 to 300 mL of gruel a day. Hansen et al. (1989) further investigated viscosity as a factor in ease of consumption for weanling children.

Adding germinated (malted) barley grain to the gruel reduced viscosity without diluting nutrient concentration. The optimum level of malt used was 1% of the barley meal. The energy density of whole-meal gruel was almost twice as high as that of gruel made from refined flour at an acceptable viscosity. Sundberg and Falk (1994) compared three Swedish barleys, one covered, one hulless and one waxy, which were ground and made into porridge. Porridges were made with 50 g of flour to 400 g of water with salt. Products were analyzed for nutritional value and properties. Degree of resistant starch was lowest in the porridges made from waxy barley, and this barley also had the highest in vitro digestibility rate, enhancing the nutritional value in situations where caloric intake was marginal.

Infant cereals in the United States are fortified with iron at 500 mg/kg, a higher level than in Europe, and the high levels of tannins and anthocyanins in barley react with the iron to form blue-black colors. Theuer (2002) addressed this problem by using a variety of proanthocyanidin-free barley to prepare porridges with several types of iron supplements. Compared to normal barley, the proanthocyanidin-free barley yielded a porridge that did not discolor when fortified at the iron level required for U.S. infant cereals.

MALTED BARLEY PRODUCTS IN FOODS

Malted barley products are used commercially for many foods to enhance color, enzyme activity, flavor, sweetness, and nutritional quality. Hansen and Wasdovitch (2005) reviewed the wide variety of malt ingredients available for use in baked goods. Diastatic malts contain enzymes that have not been deactivated by heat and are used at a very low level (about 1%) in breads as dough conditioners. Specialty diastatic malts that have been heated so as to only partially deactivate enzymes are also used in baked products. They are used at a slightly higher level (about 3%) to impart unique flavors and crumb color. Specialty nondiastatic malts are heated enough to be termed roasted. They provide warm colors and rich flavors to products without enzyme activity. Malts with various degrees of color and flavor intensity are available for specific needs. Hansen and Wasdovitch (2005) provided a table with usage rate recommendations for malt flavors and sweeteners. Malted barley extracts are produced in liquid or dry form from pure malt. Malt extracts contain about 88% total carbohydrate and 8% protein. Malt syrup, a liquid form of malt extract, has high viscosity, with 80% solids. This form of malt is a convenience for certain special products. Another well-known malt ingredient is malted milk powder. This product is a blend of malted barley, wheat extract, milk, and salt. In addition to old-fashioned malted milk shakes, malted milk powder is used for candy, ice cream, and other confectionaries. Bhatty (1996) outlined the use of hulless barley for producing malt. The benefits include a shorter steep time, direct application to food without the preparation step of malt extract or syrup, and freedom from hulls.

MISCELLANEOUS BARLEY PRODUCTS

Barley water is a beverage known since ancient times as a folk remedy or preventative for numerous ailments, such as gastroenteritis and heat exhaustion. *The Rumford Complete Cook Book* (Wallace 1930) lists barley water under "Recipes for the Sick." Currently, anecdotal records report the value of barley water for prevention or relief of side effects of chemotherapy, including nausea and mouth sores. Although no clinical reports on the benefits of barley water are available, numerous testimonials in popular Internet communications have reported its use. Barley tea made from roasted barley is commonly consumed in Asian countries. Coffee substitutes often include barley as a component. Ordinarily, barley grain, sometimes in the malted form, is roasted and combined with other plant materials, such as chicory or carob. Pazola and Cieslak (1979) investigated carbohydrate changes in the roasting process of various raw materials for coffee substitutes. Barley grain, with the highest content of dextrin and reducing sugar, produced the best quality beverage.

Barley candy is an old-time Victorian-era traditional sweet. It is a clear, hard candy made with barley water, sugar, and flavoring, and was sometimes formed into various toylike shapes. The original barley candies were reputed to have a soothing effect for throat and digestive irritations. Today, there are many "barley" candies that are no longer made with barley water. One currently available source of true barley candy is Timberlake Candies of Madison, New Hampshire.

SUMMARY

During the past 20 years, research on barley foods has seemingly exploded. Other occurrences in that period include greater awareness of health and diet relationships, awareness of cholesterol as a risk factor for heart disease, progress in knowledge about dietary fiber, food labeling, health claims for foods high in soluble fiber such as oats and barley, promotion of whole grains, increase in diabetes, concern about glycemic response—and the list goes on. We are approaching a point in nutritional information and food technology where we are able to provide consumers with acceptable, palatable foods that are synonymous with good health. Not all consumers will choose these products, but the technology exists for making healthy choices available.

Barley has a history of being a staple food in many civilizations. Its revival hinges on recognition and acceptance of its health-enhancing components, particularly β-glucan. One of the reasons that barley fell from favor as a cereal food grain is due precisely to that component, because wheat makes better bread. Much of the cereal science research reported in this chapter has been directed toward developing improved barley bread, both loaf and flat breads. Pasta is another widely popular food that can be improved nutritionally by inclusion of barley. In Chapter 5 we covered milling and processing to provide β-glucan-rich fractions to utilize in food products, and much of the research reported utilizes those ingredients. Extracted, purified β-glucan concentrates can also have a role in products as diverse as cheese, yogurt, or edible food films. Barley foods have been noted historically for health-promoting benefits, as reviewed in Chapter 1. It is appropriate that this grain, underutilized in recent years, regain its status as a prized food for good nutrition and health.

REFERENCES

AACC (American Association of Cereal Chemists International.) 2000. *Approved Methods of the American Association of Cereal Chemists*, 10th ed. AAAC, St. Paul, Minnesota.

Alford, J., and Duguid, N. 1995. *Flatbreads and Flavors: A Baker's Atlas*. William Morrow, New York.

Ames, N., Rhymer, C., Rossnagel, B., Therrien, M., Ryland, D., Dua, S., and Ross, K. 2006. Utilization of diverse hulless barley properties to maximize food product quality. *Cereal Foods World*. 51:23–28.

Andersson, A. A. M., Armö, E., Granger, E., Fredriksson, H., Andersson, R., and Åman, P. 2004. Molecular weight and structure units of (1-3), (1-4)-β-D-glucans in dough and bread made from hull-less barley milling fractions. *J. Cereal Sci.* 40: 195–204.

Anjum, F. M., Ali, A., and Chaudhry, N. M. 1991. Fatty acids, mineral composition and functional (bread and chapati) properties of high protein and high lysine barley lines. *J. Sci. Food Agric.* 55:511–519.

AOAC (American Association of Analytical Chemists International). 2000. *Official Methods of Analysis of AOAC International*, vol. II, 17th ed. W. Horwitz, ed. AOAC, Gaithersburg, Maryland.

Baik, B.-K., and Czuchajowska, Z. 1997. Barley in udon noodles. *Food Sci.Technol. Int.* 3:1–12.

Başman, A., and Köksel, H. 1999. Properties and composition of Turkish flat bread (bazlama) supplemented with barley flour and wheat bran. *Cereal Chem.* 76: 506–511.

Başman, A., and Köksel, H. 2001. Effects of barley flour and wheat bran supplementation on the properties and composition of Turkish flat bread, yufka. *Eur. Food Res.Technol.* 212:198–202.

Berglund, P. T., Fastnaught, C. E., and Holm, E. T. 1992. Food uses of waxy hull-less barley. *Cereal Foods World* 37:707–715.

Bhatty, R. S. 1986. Physiochemical and functional (bread baking) properties of hull-less barley fractions. *Cereal Chem.* 63:31–35.

Bhatty, R. S. 1996. Production of food malt from hull-less barley. *Cereal Chem.* 73:75–80.

Botero-Omary, M., Frost, D. J., Arndt, E., Adhikari, K., and Lewis, D. S. 2004. Sensory quality of chocolate chip cookies enriched with a high soluble fiber barley flour. Poster 249. American Association of Cereal Chemists, St. Paul, Minnesota.

Brennan, C. S., and Cleary, L. J. 2005. The potential use of cereal (1→3, 1→4)-β-D-glucans as functional food ingredients. *J. Cereal Sci.* 42:1–13.

Brennan, C. S., and Cleary, L. J. 2007. Utilisation of Glucagel® in the β-glucan enrichment of breads: a physicochemical and nutritional evaluation. *Food Res. Int.* 40: 291–296.

Brennan, C. S., and Tudorica, C. M. 2008. Carbohydrate-based fat replacers in the modification of the rheological, textural and sensory quality of yoghurt: comparative study of the utilization of barley beta-glucan, guar gum and inulin. *Int. J. Food Sci. Technol.* 43:824-833.

Change, K. J., and Lee, S. R. 1974. Development of composite flours and their products utilizing domestic raw materials: IV. Textural characteristics of noodles made of composite flours based on barley and sweet potatoes. *Korean J. Food Sci. Technol.* 6:65–66.

Chaudhary, V. K., and Weber, F. E. 1990. Barley bran flour evaluated as dietary fiber ingredient in wheat bread. *Cereal Foods World* 35:560–562.

Cleary, L., and Brennan, C. 2006. The influence of a (1–3) (1–4)-β-D-glucan rich fraction from barley on the physicochemical properties and in vitro reducing sugars release of durum wheat pasta. *Int. J. Food Sci.Technol.* 41:910–918.

Cleary, L. J., Andersson, R., and Brennan, C. S. 2006. The behaviour and susceptibility to degradation of high and low molecular weight barley β-glucan in wheat bread during baking and in vitro digestion. *Food Chem.* 102:889–897.

Courtin, C. M., Gelders, G. G., and Delcour, J. A. 2001. Use of two endoxylanases with different substrate selectivity for understanding arabinoxylan functionality in wheat flour breadmaking. *Cereal Chem.* 78:564–571.

Czubaszek, A., Subda, H., and Kardine-Skaradzinska, Z. 2005. Milling and baking value of several spring and winter barley cultivars. *Acta Sci. Pol. Technol. Aliment.* 4:53–62.

Dawson, K. R., Newman, R. K., and O'Palka, J. 1985. Effect of bleaching and defatting on barley distillers grain used in muffins. *Cereal Res. Commun.* 13:387–391.

Dexter, J. E., Izydorczyk, M. S., Marchylo, B. A., and Schlichting, L. M. 2005. Texture and colour of pasta containing mill fractions from hull-less barley genotypes with variable content of amylose and fibre. Pages 488–493 in: *Using Cereal Science and Technology for the Benefit of Consumers: Proc. 12th International ICC Cereal and Bread Congress.* S. P. Cauvain, S. S Salmon, and L. S. Young, eds. Woodhead Publishing, Cambridge, UK/CRC Press, Boca Raton, Florida.

Dhingra, S., and Jood, S. 2004. Effect of flour blending on functional baking and organoleptic characteristics of bread. *Int. J. Food Sci. Technol.* 39:213–222.

Dreese, P. C., and Hoseney, R. C. 1982. Baking properties of the bran fraction from brewer's spent grains. *Cereal Chem.* 59:89–91.

Edney, M. J., Rossnagel, B. G., Endo, Y., Ozawa, S., and Brophy, M. 2002. Pearling quality of Canadian barley varieties and their potential use as rice extenders. *J.Cereal Sci.* 36:295–305.

Eidet, I. E., Newman, R. K., Gras, P. W., and Lund, R. E. 1984. Making quick breads with barley distillers' dried grain flour. *Baker's Dig. Sept.* 11:14–17.

Ereifej, K. I., Al-Mahasneh, M. A., and Rababah, T. M. 2006. Effect of barley flour on quality of balady bread. *Int. J. Food Prop.* 9:39–49.

Erkan, H., Çelik, S., Bilgi, B., and Köksel, H. 2006. A new approach for the utilization of barley in food products: barley tarhana. *Food Chem.* 97:12–18.

FDA (U.S. Food and Drug Administration.) 2006. Food labeling: health claims; soluble dietary fiber from certain foods and coronary heart disease. *Fed. Reg.* 71(98): 29248–29250.

Gaosong, J., and Vasanthan, T. 2000. Effect of extrusion cooking on the primary structure and water solubility of β-glucan from regular and waxy barley. *Cereal Chem.* 77:396–400.

Gill, S., Vasanthan, T., Ooraikul, B., and Rossnagel, B. 2002. Wheat bread quality as influenced by the substitution of waxy and regular barley flours in their native and cooked forms. *J. Cereal Sci.* 36:239–251.

Gujral, H. S., and Gaur, S. 2005. Instrumental texture of chapati as affected by barley flour, glycerol monostearate and sodium chloride. *Int. J. Food Prop.* 8:377–385.

Gujral, H. S., and Pathak, A. 2002. Effect of composite flours and additives on the texture of chapati. *J. Food Eng.* 55:173–179.

Gujral, H. S., Gaur, S., and Rosell, C.M. 2003. Effect of barley flour, wet gluten and ascorbic acid on bread crumb texture. *Food Sci.Technol. Int.* 9:17–21.

Guo, G., Jackson, D. S., Graybosch, R. A., and Parkhurst, A. M. 2003. Wheat tortilla quality: impact of amylose content adjustments using waxy wheat flour. *Cereal Chem.* 80:427–436.

Hallfrisch, J., Scholfield, D. J., and Behall, K. M. 2003. Physiological responses of men and women to barley and oat extracts (Nu-trimX): II. Comparison of glucose and insulin responses. *Cereal Chem.* 80:80–83.

Han, X. 1996. Noodle-making quality from Australian Standard White Wheat and Montana barleys. M.S. thesis. Montana State University, Bozeman, Montana.

Hansen, B., and Wasdovitch, B. 2005. Malt ingredients in baked goods. *Cereal Foods World* 50:18–22.

Hansen, M., Pedersen. B., Munck, L., and Eggum, B. O. 1989. Weaning foods with improved energy and nutrient density prepared from germinated cereals: 1. Preparation and dietary bulk of gruels based on barley. *Food Nutri. Bull.* 11:40–45.

Harlan, H. V. 1925. Barley: culture, uses and varieties. *U.S. Dept. Agric. Farm. Bull. No. 1464*.

Hasler, C. M. 1998. Functional foods: their role in disease prevention and health promotion. *Food Technol.* 52:57–62.

Hatcher, D. W., Symons, S. G., and Kruger, J. E. 1999. Measurement of time-dependent appearance of discolored spots in alkaline noodles by image analysis. *Cereal Chem.* 76:189–194.

Hatcher, D. W., Lagassé, S., Dexter, J. E., Rossnagel, B., and Izydorczyk, M. S. 2005.Quality characteristics of yellow alkaline noodles enriched with hull-less barley flour. *Cereal Chemi.* 82:60–69.

Hecker, K. D., Meier, M. L., Newman, R. K., and Newman, C. W. 1998. Barley β-glucan is effective as a hypocholesterolaemic ingredient in foods. *J. Sci. Food Agric.* 77:179–183.

Hoseney, R. C. 1992. *Principles of Cereal Science and Technology*. American Association of Cereal Chemists, St. Paul, Minnesota.

Hudson, C. A., Chiu, M. M., and Knuckles, B. E. 1992. Development and characteristics of high-fiber muffins with oat bran, rice bran, or barley fiber fractions. *Cereal Foods World* 37:373–378.

IFT (Institute of Food Technologists). 2005. Functional foods: opportunities and challenges. IFT Expert Report, IFT, Chicago. Published online at http:// www.ift.org.

Ikegami, S., Tomita, M., Honda, S., Yamaguchi, M., Mizukawa, R., Suzuki, Y., Ishii, K.,
Ohsawa, S., Kiyooka, N., Higuchi, M., and Kobayashi, S. 1996. Effect of boiled
barley-rice-feeding in hypercholesterolemic and normolipemic subjects. *Plant Foods
Hum. Nutr.* 49:317–328.

Inglett, G. E., Carriere, C. J., Maneepun, S., and Tungtrakul, P. 2004. A soluble fibre gel
produced from rice bran and barley flour as a fat replacer in Asian foods. *Int. J. Food
Sci. Technol.* 39:1–10.

Irakli, M., Biliaderis, C. G., Izydorczyk, M. S., and Papadoyannis, I. N. 2004. Isolation, structural features and rheological properties of water-extractable β-glucans from
different Greek barley cultivars. *J. Sci. Food Agric.* 84:1170–1178.

Izydorczyk, M. S., and MacGregor, A. W. 1998. Rheological properties of barley nonstarch polysaccharides. Abstract. Plant Polysaccharide Symposium, University of California, Davis, California.

Izydorczyk, M. S., Storsley, J., Labossiere, D., MacGregor, A. W., and Rossnagel, B. G.
2000. Variation in total and soluble β-glucan content in hulless barley: effects of thermal,
physical, and enzymic treatment. *J. Agric. Food Chem.* 48:982–989.

Izydorczyk, M. S., Hussain, A., and MacGregor, A. W., 2001. Effect of barley and barley
components on rheological properties of wheat dough. *J. Cereal Sci.* 34:251–260.

Izydorczyk, M. S., Lagassé, S. L., Hatcher, J. E., and Rossnagel, B. G. 2005. The enrichment of Asian noodles with fiber-rich fractions derived from roller milling of hull-less
barley. *J. Sci. Food Agric.* 85:2094–2104.

Jacobs, M. S., Izydorczyk, M. S., Preston, K. R., and Dexter, J. E. 2008. Evaluation of
baking procedures for incorporation of barley roller milling fractions containing high
levels of dietary fibre into bread. *J. Sci. Food Agric.* 88:558–568.

Karaoğlu, M. M., and Kotancilar, H. G. 2006. Kavut, a traditional Turkish cereal product: production method and some chemical and sensorial properties. *Int. J. Food Sci.
Technol.* 41:233–241.

Kawka, A., Abdalla, M., and Gasiorowski, H. 1992. Barley as raw material for flat bread.
Presented at the Barley for Food and Malt Symposium, Sept. 7–9. Swedish Agricultural
University, Uppsala, Sweden.

Kawka, A., Górecka, and Gosiorowski, H. G. 1999. The effects of commercial barley
flakes on dough characteristic and bread composition. *Electron. J. Pol. Agric. Food Sci.
Technol.* 2(2). Published online at http://www.ejpau.media.pl/series/volume2/issue2/
food/art-01.html.

Kim, H.-S., and Lee, H.-J. 1977. Development of composite flours and their products
utilizing domestic raw materials: IV. Effect of additives on the bread-making quality
with composite flours. *Korean J. Food Sci.Technol.* 9:106–107.

Kim, Y. S., Ahm, S. B., Lee, K., and Lee, S. R. 1973. Development of composite flours and
their products utilizing domestic raw materials: III. Noodle-making and cookie-making
tests with composite flours. *Korean J. Food Sci.Technol.* 5:25–32.

Klamczynski, A., and Czuchajowska, Z. 1999. Quality of flours from waxy and nonwaxy
barley for production of baked products. *Cereal Chem.* 76:530–535.

Klamczynski, A., Baik, B.-K., and Czuchajowska, Z. 1998. Composition, microstructure,
water imbibition, and thermal properties of abraded barley. *Cereal Chem.* 75:677–685.

Knuckles, B. E., and Chiu, M.-C. M. 1999. β-Glucanase activity and molecular weight of β-glucans in barley after various treatments. *Cereal Chem.* 76:92–95.

Knuckles, B. E., Hudson, C. A., Chiu, M. M., and Sayre, R. N. 1997. Effect of β-glucan barley fractions in high-fiber bread and pasta. *Cereal Foods World* 42:94–99.

Köksel, H., Edney, M. J., and Özkaya, B. 1999. Barley bulgur: effect of processing and cooking on chemical composition. *J.Cereal Sci.* 29:185–190.

Lagassé, S. L., Hatcher, D. W., Dexter, J. E., Rossnagel, B. G., and Izydorczyk, M. S. 2006. Quality characteristics of fresh and dried white salted noodles enriched with flour from hull-less barley genotypes of diverse amylose content. *Cereal Chem.* 83:202–210.

Lazaridou, A., and Biliaderis, C. G. 2007. Molecular aspects of cereal β-glucan functionality: physical properties, technological applications and physiological effects. *J. Cereal Sci.* 46:101–118.

Levine, M. C., and Newman. R. K. 1986. Composition and functional characteristics of barley distillers dried grain in sausage. *Lebensm-.Wiss.Technol.* 19:198–201.

Linko, P., Harkanen, H., and Linko, Y. Y. 1984. Effect of sodium chloride in the processing of bread baked from wheat, rye and barley flours. *J. Cereal Sci.* 2:53–62.

Magnus, E. M., Fjell, K. M., and Steinsholt, K. 1987. Barley flour in Norwegian wheat bread. Pages 377–384 in: I. D. Morton-Ellis, ed. *Cereals in a European Context*. Ellis Horwood, Chichester, UK.

Mann, G., Leyne, E., Li, Z., and Morell, M. K. 2005. Effects of a novel barley, Himalaya 292 on rheological and breadmaking properties of wheat and barley doughs. *Cereal Chem.* 82:626–632.

Marconi, E., Graziano, M., and Cubadde, R. 2000. Composition and utilization of barley pearling by-products for making functional pastas rich in dietary fiber and β-glucans. *Cereal Chem.* 77:133–139.

Marklinder, I., and Johansson, L. 1995. Sourdough: fermentation of barley flours with varied content of mixed-linked (1→3),(1→4) β-D-glucans. *Food Microbiol.* 12:363–371.

McGuire, C. F. 1984. Barley flour quality as estimated by soft wheat testing procedure. *Cereal Res. Commun.* 12:53–58.

Morgan, K. R., and Ofman, D. J. 1998. Glucagel, a gelling β-glucan from barley. *Cereal Chem.* 75:879–881.

Morin, L. A., Temelli, F., and McMullen, L. 2004. Interactions between meat proteins and barley (*Hordeum* spp.) β-glucan within a reduced-fat breakfast sausage system. *Meat Sci.* 68:419–430.

Morita, N., Ando, H., Shimizu, S., Hayashi, S., and Mitsunaga, T. 1996. An application of barley flour substitution on breadmaking. Poster 30, page 89 in: *Proc. 5th International Oat Conference and 7th International Barley Genetics Symposium*, University of Saskatchewan, SK, Canada, Saskatoon, July 30–Aug. 6.

Newman, R. K., McGuire, C. F., and Newman, C. W. 1990. Composition and muffin baking characteristics of flours from four barley cultivars. *Cereal Foods World* 35:563–566.

Newman, R. K., Newman, C. W., McGuire, C. F., and Han, X. 1994. Preparation of Chinese steamed bread with hard white winter wheat and waxy barley flours. Pages 118–121 in: *Proc. International Symposium and Exhibition: New Approaches in the Production of Foodstuffs and Intermediate Products from Cereal Grains and Oilseeds*. X. Guifang and M. Zhongdeng, eds. Chinese Cereals and Oils Association (CCOA),

International Association for Cereal Science and Technology (ICC), American Association of Cereal Chemists (AACC). Beijing, China.

Newman, R. K., Ore, K. C., Abbott, J., and Newman, C. W. 1998. Fiber enrichment of baked products with a barley milling fraction. *Cereal Foods World* 43:23–25.

Niffenegger, E. V. 1964. Chemical and physical characteristics of barley flour as related to its use in baked products. M.S. thesis. Montana State University, Bozeman, Montana.

Pazola, Z., and Cieslak, J. 1979. Changes in carbohydrates during the production of coffee substitute extract, especially in the roasting process. *Food Chem.* 4:41–52.

Pedersen, B., Hansen, M., Munck, L., and Eggum, B. O. 1989. Weaning foods with improved energy and nutrient density prepared from germinated cereals: 2. Nutritional evaluation of gruels based on barley. *Food Nutr. Bull.* 11(2):46–52.

Prentice, N., and D'Appolonia, B. L. 1977. High-fiber bread containing brewer's spent grain. *Cereal Chem.* 54:1084–1095.

Qarooni, J. 1996. *Flat Bread Technology*. Chapman & Hall, New York.

Qarooni, J., Ponte, J. G., and Posner, S. 1992. Flat breads of the world. *Cereal Foods World* 37:863–865.

Quinde, Z., Ullrich, S. E., and Baik. B.-K. 2004. Genotypic variation in color and discoloration potential of barley-based food products. *Cereal Chem.* 81:752–758.

Ragaee, S., and Abdel-Aal, E. M. 2006. Pasting properties of starch and protein in selected cereals and quality of their food products. *Food Chem.* 95:9–18.

Robertson, J. A., Majsak-Newman, G., Ring, S. G., and Selvendran, R. R. 1996. Solubilisation of mixed linkage (1→3), (1→4) β-D-glucans from barley: effects of cooking and digestion. *J. Cereal Sci.* 25:275–283.

Serna-Saldivar, S. O., Guajardo-Flores, S., and Viesca-Rios, R. 2004. Potential of triticale as a substitute for wheat in flour tortilla production. *Cereal Chem.* 8:220–225.

Sood, K., Dhaliwas, Y. S., Kalia, M., and Sharma, H. R. 1992. Utilization of hulless barley in *chapati* making. *J. Food Sci. Technol.* 29:316–317.

Suhendro, E. L., Waniska, R. D., and Rooney, L. W. 1993. Effects of added proteins in wheat tortillas. *Cereal Chem.* 70:412–416.

Sundberg, B., and Falk, H. 1994. Composition and properties of bread and porridge prepared from different types of barley flour. *Am. J. Clin. Nutr.* (Suppl.): 780S.

Swanson, R. B., and Penfield, M. P. 1988. Barley flour level and salt level selection for a whole grain bread formula. *J. Food Sci.* 53:896–901.

Symons, L. J., and Brennan, C. S. 2004a. The effect of barley β-glucan fiber fractions on starch gelatinization and pasting characteristics. *J. Food Sci.* 69:257–261.

Symons, L. J., and Brennan, C. S. 2004b. The influence of (1-3)-(1-4)-β-D-glucan-rich fractions from barley on the physiochemical properties and in vitro reducing sugar release of white wheat breads. *J. Food Sci.* 69:463–467.

Tejinder, S. 2003. Preparation of characterization of films using barley and oat beta-glucan extracts. *Cereal Chem.* 80:728–731.

Theuer, R. 2002. Effect of iron on the color of barley and other cereal porridges. *J. Food Sci.* 67:1208–1211.

Torres, P. I., Ramirez-Wong, B., Serna-Saldivar, S. O., and Rooney, L. W. 1993. Effect of sorghum flour addition on the characteristics of wheat flour tortillas. *Cereal Chem.* 70:8–13.

Trogh, I., Courtin, C. M., Andersson, A. A. M., Åman, P., Sørensen, J. F., and Delcour, J. A. 2004a. The combined use of hull-less barley flour and xylanase as a strategy for wheat/hull-less barley flour breads with increased arabinoxylan and (1-3), (1-4)-β-D-glucans levels. *J. Cereal Sci.* 40:257–267.

Trogh, I., Sørensen, J. F., Courtin, C. M., and Delcour, J. A. 2004b. Impact of inhibition sensitivity on endoxylanase functionality in wheat flour breadmaking. *J. Agric. Food Chem.* 52:4296–4302.

Trogh, I., Courtin, C. M., Goesaert, H., Delcour, J. A., Andersson, A. A. M., Åman, P., Fredriksson, H., Pyle, D. L., and Sørensen, J. F. 2005. From hull-less barley and wheat to soluble dietary fiber-enriched bread. *Cereal Foods World* 50:253–260.

Tudorica, C. M., Jones, T. E. R., Kuri, V., and Brennan, C. S. 2004. The effects of refined barley β-glucan on the physico-structural properties of low-fat dairy products: curd yield, microstructure, texture and rheology. *J. Sci. Food Agric.* 84:1159–1169.

Vaikousi, H., and Biliaderis, C. G. 2005. Processing and formulation effects on rheological behavior of barley β-glucan aqueous dispersions. *Food Chem.* 91:505–516.

Vaikousi, H., Biliaderis, C. G., and Izydorczyk, M. S. 2004. Solution flow behavior and gelling properties of water-soluble barley β-glucans varying in molecular size. *J. Cereal Sci.* 39:119–137.

Wallace, L. H. 1930. Page 222 in: *The Rumford Complete Cook Book*. Rumford Co., Rumford, Rhode Island.

Wood, P. J. 2002. Relationships between solution properties of cereal β-glucans and physiological effects: a review. *Trends Food Sci. Technol.* 13:313–320.

Wood, P. J. 2007. Cereal β-glucans in diet and health. *J. Cereal Sci.* 46:230–238.

Yang, S.-R., Leem, S.-H., and Lee, W.-J. 1999. *Food Barley and Sorghum Markets in the Republic of Korea*. Research Report for U.S. Grains Council. Institute of Natural Resources, Korea University, Seoul, Korea.

8 Health Benefits of Barley Foods

INTRODUCTION

The relationship between diet and coronary heart disease has been documented extensively (Grundy and Denke 1990). As early as 1913 it was reported that feeding cholesterol to rabbits caused atherosclerosis, although this finding was not applied to human health until the 1950s. Earlier studies on diet and heart disease were focused on the effects of various types of fatty acids on serum cholesterol (Keys et al. 1957; Hegsted et al. 1965). In a landmark study covering seven countries, Keys (1970) correlated dietary intakes of saturated fats with serum cholesterol levels. In these early studies, carbohydrates in the diet were considered to be neutral in their effect on blood lipids except for causing an increase in blood triglycerides, which may be related genetically (Sane and Nikkila 1988) or specific for consumption of sugars (Macdonald 1965; Grundy and Denke 1990). McIntosh et al. (1995) reviewed the influence of barley foods on cholesterol metabolism and implications for future barley food product development.

Hugh Trowell stated that "prolonged consumption of fibre-depleted starch is conducive to the development of diabetes mellitus in susceptible genotypes" (Trowell 1973). Trowell referred particularly to populations that consume polished white rice as a staple carbohydrate. Ajgaonker had earlier cited ancient Ayurvedic physicians in India, who treated diabetes by substituting barley for rice in the diet (Ajgaonker 1972). In recent years there has been considerable research involving variations in blood glucose response to carbohydrates from different sources. The *glycemic index* (GI) is a categorization of foods based on mean blood glucose responses of a given amount of a food under closely standardized conditions (Wolever 1990) and *glycemic load* refers to quantity of the food (Salmeron et al. 1997). David Jenkins showed that rises in postprandial blood glucose and insulin levels were reduced after meals containing viscous soluble fibers (Jenkins et al. 1998). Long accepted in Europe, Canada, and Australia, the GI has only recently been recognized as significant in the U.S. food industry. The AACC International recently addressed the pros and cons of the GI (Brand-Miller

2007; DeVries 2007; Jenkins 2007; Jones 2007). More recent approaches to the basis of the glycemic effect of barley are the amylose/amylopectin ratios (Behall et al. 1989) and the presence of resistant starch due to the processing of amylose (Topping and Clifton 2001). Pins and Kaur (2006) reviewed the effects of barley β-glucan on the risk of both cardiovascular disease and diabetes, stressing the importance of lifestyle changes such as diet before initiating drug therapy.

In this chapter we review the healthful benefits of barley, focusing on blood lipids and glycemic responses. Other health benefits, such as gastrointestinal health, are also discussed briefly. It is important to note that the cutting edge of nutritional research delves into underlying mechanisms and biological controls, and future studies will more specifically provide answers to today's questions regarding the "how" of physiological responses.

BARLEY AND HEART DISEASE

It has been well established that elevated serum cholesterol is a risk factor for coronary artery disease, which remains a major cause of chronic illness and continues to be the leading cause of death in the United States (NCEP, 2001; CDC, 2007). Trowell (1975) promoted the hypothesis that dietary fiber may play an important role in regulating serum cholesterol; however, at that time, investigations centered on wheat fiber. Rimm et al. (1996) conducted a prospective study of over 50,000 male health professionals and concluded that there was an inverse relationship between fiber intake and myocardial infarction. Within the three main food contributors to total fiber intake (i.e., vegetables, fruit, and cereals), cereal fiber was most strongly associated with the prevention of heart disease. DeGroot et al. (1963) had earlier reported the cholesterol-lowering effect of rolled oats, and throughout the 1980s and 1990s a large number of research studies followed this trend, as summarized by Truswell (2002).

The cholesterol-reducing property of barley was shown initially in chickens by Fisher and Griminger (1967) and later confirmed by Fadel et al. (1987), Martinez et al. (1992), and Bengtsson et al. (1990). Danielson et al. (1997) reported on the hypocholesterolemic effect of barley milling fractions as related to the viscosity of the gut digesta of chickens, and Mori (1990) demonstrated the relationship of barley soluble fiber and fecal excretion of lipids with consequent lowering of blood cholesterol in rats. Kalra and Jood (2000) fed diets made from three barley cultivars to rats and compared them with a casein control diet. Dolma, the barley with the highest β-glucan content (5.39%), produced the greatest reduction in total and LDL cholesterol. Bird et al. (2004a) reported reduced serum cholesterol in young pigs fed a high-amylose barley cultivar with high β-glucan content. One long-term study (Li et al. 2003a) showed that barley diets in rats produced sustained depression of plasma total cholesterol, triglycerides, and free fatty acids over nine months. A recent Japanese study (Araki et al. 2007) reported reduced accumulation of abdominal adipose tissue in rats fed diets with high-β-glucan barley, as well as reduced size of adipose cells, compared with animals fed wheat diets.

Mechanisms Responsible for Cholesterol Effects

As reports of cholesterol-lowering effects of both oats and barley increased, speculation about the causative mechanisms occurred among scientists. One early consideration concerned the intestinal viscosity effect of β-glucan, which is believed to increase the thickness of the unstirred layer of the small intestine, slowing and inhibiting the absorption of lipids and cholesterol (Wang et al. 1992). The cholesterol-lowering property of oats is attributed to viscosity formed in the intestine due to β-glucans (Tietyan et al. 1990). Oat and barley β-glucans are basically the same, varying primarily in the concentration and degree of solubility and molecular weight among varieties (Wood 1986). Delaney et al. (2003) reported increased fecal neutral sterol concentrations and reduced aortic cholesterol levels in hamsters fed either oats or barley, and concluded that the two grains were equal in cholesterol-lowering effect. Arabinoxylans are also capable of producing viscous solutions and hence contribute somewhat to the hypocholesterolemic effect of both grains. Quantitatively, some barley varieties are known to have a wide range in β-glucan levels, as described in Chapter 3.

β-Glucans are also believed to cause binding of bile acids in the intestine, causing them to be excreted in fecal waste, with the result of body cholesterol being broken down to replace them. Thus, changes in bile acid metabolism in response to β-glucan have been implicated in their hypocholesterolemic action (Kahlon et al. 1993). Jenkins et al. (1993) conducted a crossover study with hyperlipidemic men and women on a low-fat diet, in which foods with either soluble or insoluble fiber from various sources were consumed. When soluble fiber was consumed, subjects' LDL cholesterol was the lowest and fecal bile acid excretion was the greatest. In an in vitro study, Kahlon and Woodruff (2003) reported that dehulled barley and β-glucan-enriched barley both exhibited bile acid binding under physiological conditions. Yang et al. (2003) fed diets with 30% waxy barley to cholesterol-fed rats. There was greater bile acid excretion in animals fed barley compared to controls, and serum total and LDL cholesterol were decreased 19 and 24% respectively. In addition, activity of cholesterol 7α-hydroxylase was up-regulated 2.3-fold in barley-fed rats, indicating increased synthesis of bile acids to replace those excreted.

Human Clinical Studies

Barley has been shown to be equally effective in human lipid metabolism trials as in animal studies (Behall and Hallfrisch 2006). Because wheat is a commonly used grain in baked products and differs from barley in the proportion of soluble fiber, it is generally used as a control product. Newman et al. (1989a) compared a variety of baked products and cereal foods made from barley or wheat in a clinical study to measure possible effects on plasma cholesterol levels. Fourteen healthy volunteers consumed three products containing either barley or wheat daily for four weeks. Subjects who consumed wheat foods had small but significant increases in total and LDL cholesterol compared with pretreatment levels, whereas those who consumed barley had significantly lower cholesterol levels. In a subsequent study with 22 hypercholesterolemic individuals, this same research group compared muffins, flatbread, nugget cereal, and cookies made with barley or oat flour. For both oat and barley foods, there was significant reduction in serum total and LDL cholesterol from base levels, with no difference between the two groups. The latter study (Newman et al. 1989b) clearly demonstrated that barley is equivalent to oats in lowering cholesterol.

McIntosh et al. (1991) conducted a crossover study of 21 hypercholesterolemic men, in which bread, muesli, spaghetti, and biscuits made from either barley or wheat were consumed for four weeks. Total plasma cholesterol and LDL cholesterol were lowered significantly in subjects who consumed barley products compared to wheat products. A remarkable study from Japan (Ikegami et al. 1996) demonstrated a strategy for dietary variation to increase soluble fiber intake in a large population. This research combined cooked white rice and pearled barley in a 50 : 50 mix which was consumed by 20 hypercholesterolemic men at each meal for four weeks. Because rice is a staple food in Japan, subjects were provided convenient packages of the precooked barley–rice blend, to be microwaved and consumed with meals instead of their customary plain white rice. The researchers reported a significant drop in total and LDL cholesterol from baseline levels. A second phase of this study involved female subjects, and comparable results were achieved. In a similar study conducted later in Japan (Li et al. 2003b), 10 women subjects consumed either 100% rice or 30% barley with 70% rice three times daily for four weeks. The barley intake reduced plasma total and LDL cholesterol significantly. A third study from Japan conducted by Shimizu et al. (2007) investigated whether pearled barley as an alternative to rice would reduce visceral body fat as well as reduce LDL and total cholesterol in 44 hypercholesterolemic men. The pearl barley intake reduced cholesterol significantly in subjects consuming barley, and there were also reduced waist measurements and reduced visceral fat stores as measured by CT scans, all positive factors for prevention of heart disease.

Behall et al. (2004a) of the USDA Agricultural Research Service conducted a study on hypercholesterolemic men. Subjects were fed the American Heart Association step 1 diet (AHA 1988) for two weeks, followed by the same diet with 20% of the energy replaced by either (1) brown rice/whole wheat, (2) $\frac{1}{2}$ barley and $\frac{1}{2}$ brown rice/whole wheat, or (3) barley, for five weeks. Total dietary fiber in the diets was 27 to 33%, and added soluble fiber from the barley varied from 0.4 g (low) to 3 g (medium) to 6 g (high). Compared with prestudy baseline levels, total cholesterol and LDL cholesterol were significantly lower in subjects after all levels of added soluble fiber. The percent of cholesterol reduction followed a dose–response pattern, with maximum reductions of 20% for total cholesterol and 24% for LDL cholesterol. Serum triglycerides were lowered as well, and HDL cholesterol was increased, providing an additional health risk advantage for subjects. These researchers also measured lipid subclass fractions in plasma of experimental subjects. The size and density of LDL particles have been implicated as risk factors for heart disease (Freedman et al. 1998). Increased risk has been associated with a gender-related predominance of small, dense LDL particles. In the present study, the mean number of LDL particles decreased after all experimental diets, and the greatest decrease occurred after the high-soluble fiber diet. This same research group conducted a similar study with hypercholesterolemic men and women, both pre- and postmenopausal, who consumed barley (Behall et al. 2004b). Diets in this second study were calculated on the basis of β-glucan levels, containing 0, 3, and 6 g per day from barley foods. After five weeks cholesterol was significantly lower in subjects when the diet contained 3 or 6 g of β-glucan, and the greatest change occurred in the men and postmenopausal women. Contrary to the previous experiment, HDL cholesterol and triglycerides did not differ in subjects consuming any level of β-glucan. In this study, LDL particle size significantly decreased when all whole grains were incorporated into the diets. The authors concluded that the barley diets were effective in lowering heart disease risk.

Another group (Lia et al. 1995) provided bread made with β-glucan-rich barley to ileostomy subjects and reported increased cholesterol and bile acid excretion in correlation with lowered blood cholesterol. Jenkins et al. (2005) compared cholesterol-lowering foods, including barley, with a statin in 34 hyperlipidemic patients and reported no significant difference in lowering LDL cholesterol between treatments. Statins act by inhibiting the action of the rate-limiting enzyme of cholesterol synthesis, HMG-CoA reductase, in the liver. For patients who cannot or desire not to take statins, this study represents dietary modification as an alternative method of controlling serum cholesterol.

Spent brewers' grain, also referred to as barley bran flour, has been tested as a hypocholesterolemic ingredient (Chaudhary and Weber 1990). Lupton et al. (1994) supplemented diets of 79 hypercholesterolemic subjects with either 30 g per day of barley bran flour or 3 g per day of barley oil. Both supplements significantly decreased total and LDL cholesterol after 30 days. In that study, the barley oil treatment effect was attributed to tocotrienol content rather than β-glucan. Zhang et al. (1991) studied individuals with ileostomies who were given high-fiber diets containing spent brewers' grain. Subjects had increased excretion of cholesterol concurrent with lowered serum LDL cholesterol.

Extracted β-Glucan as a Food Additive

The concept of utilizing extracted, concentrated oat or barley β-glucan as an additive to foods to produce a health benefit without the total bulk and calories of the original grain has been pursued. Brennan and Cleary (2005) published an exhaustive review on the potential of cereal β-glucans as functional food ingredients, which stressed the necessity of optimal extraction procedures to assure consistent efficacy of products in which the β-glucan is incorporated. Poppitt (2007) reiterated the need for caution in assuming that β-glucan extracts are always effective in lowering cholesterol. In human studies of barley β-glucan efficacy, whether whole grain, milling fractions, or extracted β-glucan preparations are used, there can be inconsistencies in results. The original variety of barley used, the processing and analytical techniques used, and the preparation of foods to be consumed are all variables. In addition, the selection of subjects, sample size, dose of treatment, and dietary control of subjects may influence the results. Therefore, there are some inconsistencies and conflicting results reported in early clinical studies using extracted β-glucans.

Keogh et al. (2003) investigated the effects of a concentrated barley β-glucan product (Glucagel) incorporated into foods consumed by hypercholesterolemic men over four weeks. Compared to a control period, there were no significant changes in total cholesterol and other blood lipids, although subjects consumed 8 to 11.9 g of β-glucan per day. The authors concluded that structural changes might have occurred in the β-glucan during extraction or storage. Björklund et al. (2005) incorporated β-glucan fractions from oats and barley into fruit-flavored beverages to provide a daily intake of 5 or 10 g of β-glucan from either grain. A total of 89 hypercholesterolemic men and women at two sites consumed the beverages for a five-week period. Compared to controls, 5 g of β-glucan from oats significantly lowered total cholesterol, but there was no effect on serum lipids by either 10 g of oat or either level of barley β-glucan. These inconsistent results were attributed to differences in molecular weight and/or solubility of β-glucans due to processing. These two studies indicate the necessity for careful evaluation of products intended for use in functional foods. However, Naumann et al. (2006) demonstrated effective lowering of total and LDL cholesterol in healthy subjects who consumed a fruit drink enriched with oat β-glucan. Keenan et al. (2007) reported use of concentrated β-glucan extract from barley to lower LDL cholesterol in hyperlipidemic subjects. Because of prior reports on the influence of the molecular weight of β-glucan preparations on viscosity and possible non-efficacy, two products, of high and low molecular weight (HMW and LMW), were used in the study. The β-glucan extracts were tested at daily doses of 3 and 5 g, incorporated into juice and cereal. After six weeks of treatment, LDL cholesterol levels fell by 15% in those consuming 5 g of HMW, 13% in those consuming 5 g of LMW, and 9% when both groups consumed 3 g of β-glucan per day. The researchers concluded that molecular weight did not significantly influence efficacy in this study, although the LMW product had improved sensory properties and might have greater applicability in food products. In two experiments on β-glucan from oat bran (Kerckhoffs et al. 2003), there were differences in the results based on the food vehicle used to administer the products. Subjects fed bread and cookies providing 5 g of β-glucan daily for four weeks had no significant change in total or LDL cholesterol. After a washout period, the same subjects consumed 5 g of the same product per day in orange juice, and their cholesterol levels were decreased significantly. These authors concluded that the processing of baked products may have had a detrimental effect on cholesterol-lowering properties of the β-glucan, but the molecular weight did not appear to be a factor. At this point, there are insufficient data on the effects of processing to draw meaningful conclusions on the influence of processing, baking, or cooking on the efficacy of extracted β-glucan on blood lipid composition. The variability of outcomes in studies using β-glucan reinforces the importance of standardization of extraction procedures, as cited by Brennan and Cleary (2005).

GLYCEMIC RESPONSE TO CARBOHYDRATE CONSUMPTION

The worldwide incidence of diabetes is increasing rapidly, as well as a related condition called the *metabolic syndrome* (CDC 2007). Both of these conditions are related to carbohydrate metabolism, in which food carbohydrates are digested into glucose and absorbed as such into the bloodstream. This action stimulates the release of insulin from the pancreas into the blood, facilitating the entry of glucose into body cells. In diabetes, there is either an absence of insulin (type 1) or inefficient use of insulin (type 2) by the cells, causing glucose to build up in the blood, thereby creating serious complications if treatment is not provided. In the condition known as the metabolic syndrome, the body cells do not respond normally to insulin, resulting in elevated insulin as well as elevated glucose in the blood. This relative insulin resistance initiates further complications, such as abdominal obesity, hypertension, and elevated blood triglycerides.

Carbohydrate foods differ in their rate of digestion and absorption into the blood, and these differences were the basis of establishing the glycemic index (GI) (Brand-Miller 1994). Food carbohydrates, especially starch, that have slower rates of digestion, are sometimes termed "lente" or slowly digested and absorbed carbohydrates. In general, foods with a high GI value are processed rapidly in the digestive tract, causing a steep high peak in blood glucose, and in normal individuals are followed by a corresponding rapid release of insulin. Low-GI foods, on the other hand, have slower absorption with a blunted, lower, and more prolonged glucose peak. The AACC addressed the need for consistent information about glycemic response by developing standard definitions of terms (Jones 2007) that could enable food manufacturers to communicate the glycemic effect of foods to consumers.

The stomach-emptying rate is a major factor in the delivery of glucose into the small intestine for absorption (Horowitz et al. 1993). Fat and protein tend to delay gastric emptying, and some studies have concluded that the GI of individual foods may not accurately predict the glycemic and insulinemic effect of mixed-nutrient meals as foods are normally eaten. Wolever et al. (2006) demonstrated that when foods were tested in meal-type settings, the GI can reliably predict the glycemic effect of mixed meals, regardless of protein or fat content. In their study, the carbohydrate content and GI determined 90% of the variation observed in mean glycemic response to the mixed meals. Other food functional properties affecting the digestion and absorption of carbohydrates include the amount and type of fiber, food structure, physical texture, hydration of starch, and starch type (Björck et al. 1994; Björck and Elmståhl 2003; Huth et al. 2000). Foods being tested for glycemic responses are conducted on subjects in the fasting state. Measured quantities of test foods are consumed, after which blood samples are measured for glucose and insulin concentrations at timed intervals after eating. The glucose and insulin blood concentration plots over time are calculated into areas under the curve (AUC) and the time to reach peak concentrations. Graphs with superimposed curves of different products are often an effective visual comparison of responses to the foods.

The method of measuring the GI is time consuming and expensive, with unaccountable variability in repeated testing and between individuals. Some critics contend that the GI is merely an approximate prediction of glycemic response. Other physiological markers have recently been utilized as alternative methods to indicate and measure glycemic responses to foods. The term *incretin* refers to the hormones glucagon-like polypeptide (GLP-1) and glucose-dependent insulinotropic polypeptide (GIP), which are now measured in glycemic studies. Incretin hormones are involved in the interaction between glucose ingestion and insulin response (Nauck et al. 2004; Gentilcore et al. 2006; Milton et al. 2007). It is believed that dietary interventions with the potential of controlling postprandial glycemic responses are more definitive if the level of incretin hormone secretion is determined. Breath hydrogen (H_2) is frequently measured postprandially to assess colonic fermentation of indigestible carbohydrates such as soluble dietary fiber. The degree of fermentation is assessed by the quantity of H_2 expired over a specified time period (Thorburn et al. 1993). An increasing amount of breath H_2 is indicative of increased fermentation in the large intestine, also resulting in the formation of the short-chain fatty acids (SCFAs): acetic, propionic, and butyric.

Plasma free fatty acids (FFAs) are produced by lipolysis of triglyceride stores and are involved in the dynamic energy balance in body cells. Insulin and other hormones have a direct influence on serum FFA concentration, and high FFA levels tend to decrease insulin effectiveness in moving glucose into body cells. Foods with a high GI tend to result in elevation of FFAs, creating a cycle leading to insulin resistance. Hence, a lowering of FFA levels following a dietary intervention for glycemic effect would be seen as a positive benefit (Ferannini et al. 1983; Wolever et al. 1995). FFA levels are also believed to be responsive to the presence of SCFAs produced by colonic fermentation of indigestible carbohydrates. Adiponectin, a hormone that is secreted from adipose tissue, modulates glucose regulation and fatty acid catabolism. Because it has been associated with protection from metabolic conditions related to diabetes, such as obesity, adiponectin has recently been measured in glycemic studies (Kim et al. 2007). Insulin resistance in susceptible individuals (impaired glucose tolerance, obesity, or hyperlipidemia) can be detected or predicted by various indices, such as the homeostasis model assessment (HOMA) and other methods (Behall et al. 2005). Insulin resistance has been detected and measured using the euglycemic–insulinemic clamp in experimental animals, in which decreased insulin uptake by peripheral tissues is recorded (Yokayama and Shao 2006). Finally, interleukin 6 (IL-6), a pro-inflammatory cytokine, is a metabolic marker of immune response to inflammation. Elevated levels of IL-6 are associated with hepatic glucose production, and thus have implications for glycemic evaluations. Increased levels of cytokines inhibit translocation of glucose transporters to the cell surface in response to insulin (Kishimoto 1989). Future research will undoubtedly clarify the roles of these mechanisms in terms of carbohydrate metabolism for both healthy individuals and those with metabolic syndrome or diabetes.

Effect of Barley on Carbohydrate Metabolism

Investigation of the effect of barley on carbohydrate metabolism has intensified since Sato et al. (1990) reported that plasma glucose concentration in patients both with and without diabetes was lower after barley consumption than after rice consumption. Research studies on the glycemic effects of barley have focused on one of two major aspects: (1) β-glucan as a viscous fiber, and (2) starch composition ratio of amylose to amylopectin. In many cases the glycemic response to barley foods will be due to a combination of these two factors. Ikegami et al. (1991) in Japan measured glycemic responses in normal and diabetic rats fed a barley diet. All animals exhibited improved glucose tolerance, and fasting blood glucose in diabetic rats fed barley was reduced to normal levels. Similar results in diabetic rats were reported by Li et al. (2003a). Narain et al. (1992), from India, studied metabolic responses to barley in healthy human subjects. Chapatis made from barley flour were consumed in a quantity to provide 40% of the total daily cereal intake. After four weeks, the incremental area under the 3-hour glucose curve decreased from 107.9 mg/dL to 91.5 mg/dL. Battilana et al. (2001) investigated the mechanism of action of β-glucan in postprandial glucose metabolism in healthy men. These researchers concluded that the lowered glycemic response following a meal containing β-glucan is related to delayed and/or decreased absorption of glucose due to increased viscosity in the gut. Following these early studies on barley as an agent for modulation of blood glucose, together with recognition of the GI, there was heightened interest in the development of products containing barley, especially due to the growing incidence of diabetes.

The Department of Applied Nutrition and Food Chemistry at the University of Lund in Sweden has rapidly become a recognized center for research in glucose metabolism related to cereal grains. Liljeberg et al. (1996) measured postprandial blood glucose and insulin responses to cereal products made from common barley, oats, and Prowashonupana (PW) barley (later renamed Sustagrain). The latter barley contained 19.6% soluble fiber, most of which was β-glucan. Porridges made from the oats and common barley produced postprandial glucose and insulin responses similar to the white bread reference control. In contrast, porridge made from PW barley induced significantly lower responses than that of the control. Also studied were two types of flour-based bread products (flatbreads) composed of whole-meal flour made from PW and common barleys in ratios of 50 : 50 and 80 : 20, respectively. The glycemic response of these products was significantly lower than in a similar flatbread product made from white wheat flour. The high-fiber barley products were generally liked by the subjects, especially the breads. Conclusions reached in this report were that porridge and flatbreads made from PW barley were consumer-acceptable and effective in producing a lowered glycemic response.

Lifschitz et al. (2002) compared PW barley (15% β-glucan) with a standard barley cultivar (5.3% β-glucan) grown in a chamber containing radioactive carbon dioxide ($^{13}CO_2$), in order to measure the rate of CO_2 absorption in subjects who consumed the barley cooked in a whole-grain form. This study determined absorption using two indicators: (1) oxidation by means of digestion and absorption, and (2) malabsorption. For oxidation, breath CO_2 was measured as an indicator of utilization of carbon from barley. Malabsorption of the indigestible β-glucan was determined by measuring breath hydrogen. The results of this study clearly demonstrated that the absorption of carbohydrate from the PW barley was significantly lower than that of the standard barley, presumably due to the concentration of β-glucan and its attendant viscosity and delay in absorption.

An Australian group (Keogh et al. 2007) tested the high-amylose barley Himalaya 292 (renamed BarleyMax) and wheat, formulated into meals for healthy lean women in a crossover study to determine glycemic responses and satiety effects. The mean areas under the curve in response to wheat-containing meals were 22 and 32% higher than those for barley-containing meals for glucose and insulin, respectively. In this study, the barley meals did not reduce spontaneous food intake, despite its reduced glycemic and insulinemic effects.

A β-glucan concentrate, Nu-trimX, developed from an extract of PW barley, and a similar product from oats (Inglett 2000) were tested for glucose and insulin responses by Hallfrisch et al. (2003) at the USDA Human Nutrition Research Center in Beltsville, Maryland. Healthy men and women consumed test meals containing a pudding with either oat bran, barley flour, oat Nu-trimX, or barley Nu-trimX, with glucose as a control. The area under the blood glucose curve was lowest for the barley extract, but not significantly different from other β-glucan sources, which were all lower than the glucose control. These researchers concluded that extracted hydrocolloids extracted from oats and barley retain the beneficial effects of the original whole grains. Barley β-glucan extract provided lower glucose and insulin curves than resistant starch in muffins (Behall et al. (2006b). In that study there were two groups of male subjects, normal weight and overweight, and the latter exhibited higher blood glucose increments after consuming the test muffins. However, in an earlier similar study with women, β-glucan and resistant starch produced similar glycemic responses (Behall et al. 2006a). Poppitt et al. (2007) investigated the glycemic effect of a high barley supplement (Cerogen) added to solid foods and to a beverage. The supplement improved glucose control when added to solid foods, but not when consumed as a beverage. Researchers attributed the results to the lack of viscosity of the supplement in liquid form, which is indicative of enzymatic hydrolysis of the β-glucan. This phenomenon is commonly observed when barley is moistened without destroying the endo-β-glucanases.

The USDA Western Regional Center at Albany, California also conducted studies on the glycemic effects of barley foods (pasta). Using a procedure of milling and sieving (Knuckles et al. 1997), Yokayama et al. (1997) prepared a β-glucan-enriched flour from the hulless, waxy Waxbar barley, containing 32% TDF and 20.1% β-glucan. Pasta made from a blend of the barley flour and wheat flour contained 7.7% β-glucan. Healthy subjects consumed the pasta in meals containing 100 g of available carbohydrate, using an all-wheat pasta as a control. The barley pasta resulted in a lower glycemic response as measured by average total area and maximum increment of the blood glucose curves. Lower insulin responses to the barley pasta were also observed. This research group conducted a second pasta study in which two barley cultivars were compared (Bourdon et al. 1999). Pasta was prepared by replacing 40% of wheat flour with either Waxbar barley flour enriched in β-glucan as in the previous study, or

PW flour, which naturally contains a high β-glucan concentration. The resulting pastas contained approximately 5 g of β-glucan per test meal serving and were compared with all-wheat pasta for glycemic response. In this study, the objectives were to measure insulin, glucose, cholecystokinin, and lipid levels in blood after pasta meals. The two barley meals produced slower carbohydrate absorption and increased plasma triglycerides and cholecystokinin. The latter is believed to be related to delay in gastric emptying and blunted glycemic responses. More recently, Cleary and Brennan (2006) demonstrated delayed reducing sugar release from pasta containing a β-glucan-rich barley fraction.

Cavallero et al. (2002) processed whole-grain barley flour (4.6% β-glucan) into two types of concentrated β-glucan fractions, one by sieving (8.5% β-glucan) and one by water extraction (33.2% β-glucan). Using a commercial bread wheat flour, the researchers produced blends with the barley fractions for incorporation into breads. The whole barley flour and the sieved fraction were included at 50% of the bread formula and the water-extracted fraction at 20% of the formula, to provide a stepwise increased distribution of β-glucan in the products. Resulting breads contained 2.4, 4.2, and 6.3% β-glucans, compared to all-wheat bread containing 0.1% β-glucan. Breads were consumed by healthy subjects in portions containing 60 g of available carbohydrate to evaluate glycemic response. A linear decrease in calculated GI was found relative to increasing β-glucan content. These researchers also compared bread quality characteristics and concluded that the water-extracted high-fiber fraction provides an effective and palatable source of fiber enrichment for bread with a lowered glycemic effect.

Many of the glycemic studies using barley soluble fiber have mentioned the term *satiety*, which is logically related to delayed gastric emptying and slow absorption. It is also logical to relate satiety to weight control and prevention of obesity. Kim et al. (2006) measured short-term satiety response as well as glycemic response in overweight men and women fed cooked cereal (wheat and/or barley) that contained either 0, 1, or 2 g of barley β-glucan. Researchers used the visual analog scale (VAS) to assess sensory meal response as well as blood glucose after the meals. Only after the 2-g β-glucan meal was there significant reduction in blood glucose curves, and the VAS scores were not different among groups, indicating that there was no detectable difference in satiety value. An Australian group (Keogh et al. 2007) tested the high-amylose barley Himalaya 292 (BarleyMax) and wheat, formulated into meals for healthy lean women in a crossover study to determine glycemic responses and satiety effects. The mean areas under the curve in response to wheat-containing meals were 22 and 32% higher than those for barley meals for glucose and insulin, respectively. In this study, the barley meal did not reduce spontaneous food intake despite its reduced glycemic and insulinemic effects.

The Swedish group at the University of Lund have contributed important methodological improvements to research in this area. Östman et al. (2006) developed a method of predicting glycemic effect of a product based on fluidity (a measure of Newtonian viscosity) of in vitro enzymatic digesta derived from barley bread products incubated with α-amylase. Fluidity was determined with a consistometer, which is a measure of the distance a fluid flows in a given time interval. In this study, PW barley flour, previously studied by this group, was prepared into bread at different concentrations. Glycemic responses followed the trend of increasing β-glucan content, with a minimum GI value of 52.3 compared with 100 for white wheat bread. The fluidity index values were highly correlated with GI values of the entire range of breads. Interestingly, in this study the palatability of the barley test breads was preferred by the test subjects. The fluidity measurement technique offers a relatively simple tool for evaluating glycemic properties of a food enriched with viscous fiber. Earlier methodology reports by the Lund group were measurement of gastric emptying rate by ultrasonography (Garwiche et al. 2001) and use of an in vitro procedure based on chewing to predict metabolic responses (Granfeldt et al. 1992).

Studies Using Subjects with Diabetes

Shukla et al. (1991) compared the glycemic response of barley chapatis in both healthy subjects and those with type 2 diabetes. A 50-g carbohydrate serving of chapati was consumed by each subject followed by glycemic and insulinemic response measures. The GI for barley chapati was 68.7 in healthy subjects and 53.4 for those with diabetes. The researchers observed a higher insulinemic index in diabetic subjects than in healthy subjects, suggesting higher mobilization of insulin. Pick et al. (1998) incorporated waxy hulless barley flour into a variety of breads, pasta, muffins, and cookies, some of which included cracked barley grain. The subjects with type 2 diabetes consumed eight servings daily of the barley products, containing 15 g of carbohydrate per serving, for a 12-week period. Each subject served as his own control, consuming white wheat bread for an equivalent period. Mean total dietary fiber intake was 18 g per day in the control period and 39 g per day in the barley period. Mean glycemic response (AUC) was lower, and insulin response area was higher, for barley foods than for white bread. The higher postprandial serum insulin response was of particular interest, since other studies with soluble fiber had resulted in variable effects on insulin levels. The barley bread products were reported to be very palatable to subjects, and the authors also reported that some subjects reduced their dose of oral hypoglycemic drugs while in the barley period.

Rendell et al. (2005) compared breakfast meals of PW barley flakes and oatmeal for glycemic response in patients with and without diabetes. A low-fiber liquid meal replacer (LMR) was used as a reference standard. A substantial reduction in the postprandial glycemic peak following ingestion of the PW barley was observed as compared to LMR or oatmeal. Insulin response following PW was dramatically lower than either oatmeal or LMR, prompting researchers to calculate glucose/insulin ratios (G/I), which were found to be lowest for LMR and highest for PW. The G/I ratios of diabetic subjects were consistently higher than those of the nondiabetic subjects. These researchers commented that glycemic responses from PW barley were similar to those observed in diabetic patients treated with α-glucosidase, a drug used to block carbohydrate digestion and absorption. In Japan, Hinata et al. (2007) reported markedly improved diabetes metabolic control among male prisoners who were provided a high-dietary fiber diet that included boiled rice with barley in a 50 : 50 blend, similar to that advocated by Ikegami et al. (1996) for cholesterol metabolism.

Starch Amylose/Amylopectin Ratio and Resistant Starch

Both amylose and amylopectin are present in most plant starches, and the ratio varies not only between grains but between varieties within the same species. For example, several grains, such as wheat, corn, rice, and barley, exhibit waxy varieties that contain higher-than-normal proportions of amylopectin, and conversely, in some instances, higher-than-normal amylose in their starch (see Chapter 4). The amylose/amylopectin ratio of starch provides a distinct difference in glucose and insulin responses, as shown in studies with rice (Goddard et al. 1984) and cornstarch (Behall et al. 1989). Xue et al. (1996) heat-treated starch from two high-amylose (40% of amylose in the starch) barleys and demonstrated reduced starch digestibility in vitro and depressed glucose peaks when fed to animals, compared with waxy and normal starch barley. The effects were magnified when the starches were autoclaved rather than boiled, demonstrating that the cooking characteristics of a starch are affected both by the proportions of amylose and amylopectin and by the cooking method. High-amylose starches have unique filming and binding properties and have been utilized for special purposes in the food industry (Hullinger et al. 1973). Gelatinization, a swelling process that occurs in starch granules when they are heated in water, varies with starch types. Gelatinization temperature increases relative to the proportion of amylose in a starch (Zapsalis and Beck 1985).

Being highly polar linear polymers, amylose molecules show a strong tendency to associate and link together through hydrogen bonding. This causes either a firm gel with cross-bonding or an insoluble precipitation. The process is called *retrogradation*, or formation of *resistant starch* (RS), defined as that fraction of starch that is resistant to digestive enzymes and enters the large bowel of healthy humans as an intact large-molecular-weight polymer (Asp 1992). Botham et al. (1997) demonstrated physicochemical techniques for characterization of resistant starch in a human ileostomate. In the bowel (or colon or large intestine), the RS, plus soluble fiber, including β-glucan, is fermented by microflora, producing increased levels of short-chain fatty acids (SCFAs), which are reported to be beneficial in maintaining bowel health (Topping and Clifton 2001).

Flours made from a genotypic series of barleys from the cultivar Glacier, with amylose content varying from 8 to 35% (waxy, normal, and high-amylose) were milled into flour and evaluated by Björck et al. (1990) for amylolysis and gelatinization. In this study, the amylose/amylopectin ratio had little influence on gelatinization or susceptibility to α-amylolysis following boiling. However, with increasing amylose content there was a concomitant decrease in the in vitro digestibility of autoclaved flours due to retrogradation, as in the study reported by Xue et al. (1996). Correspondingly, the high-amylose barley contained the most RS. The barley flours varied in dietary fiber in decreasing order with their starch types: waxy > high-amylose > normal starch. When added to a starch suspension, isolated fiber preparations from all barleys were equally effective in reducing the rate of gastric emptying in animals. In another study from this laboratory, barley products made from four cultivar lines derived from Glacier barley, with varying amylose/amylopectin ratios, were evaluated (Granfeldt et al. 1994). Each barley genotype was tested in a boiled intact kernel form or porridge made from milled kernels. All barley products elicited lower glycemic and insulinemic responses and higher satiety scores compared with white wheat bread, and the intact kernels exhibited the lowest metabolic responses. This study included a unique in vitro measurement of starch hydrolysis based on chewed food samples incubated with pepsin to mimic physiological conditions. The high-amylose products released starch more slowly during enzymatic incubation than did the corresponding products with less amylose. The RS content of food products ranged from 0.4% in waxy to 5.6% in the high-amylose flour product. Åkerberg et al. (1998) prepared breads with the Glacier series of barleys, made under normal baking conditions (45 minutes, 200°C) or "pumpernickel" conditions (20 hours, 120°C). In both baking series, the amount of RS increased with increasing amylose content, and the long-time/low-temperature baked breads were generally higher in RS, ranging from 2% in waxy barley to 10% in high-amylose barley bread. The long-time/low-temperature bread also had a lower rate of in vitro starch hydrolysis than that of white wheat bread, and a reduced blood glucose response in subjects. Overall, the high-amylose barley bread baked under special conditions was considered desirable for nutritional starch properties.

Effects of Food Form and Processing on the Glycemic Index

The physical form of a grain at the time of consumption obviously has an effect on its physiological availability for digestion. Even its passage through the mouth can vary in the degrees of mastication and exposure of surface area to salivary enzymes. Granfeldt et al. (1992) and Liljeberg and Björck (1994) evaluated bread products containing whole meal or intact kernels for GI and insulinemic index in healthy subjects. The presence of kernels in the breads always produced lower glycemic responses, regardless of grain type, and parallel in vitro hydrolysis rate index (HI) values. The correlations of glycemic index and insulinemic index to the HI were very high, suggesting that the in vitro method could be predictive of glycemic response. Livesey et al. (1995) used human subjects with ileostomies to investigate the digestion of milled and flaked barley and to determine whether the physical form affected digestion. The subjects, all of whom had previous colon removal, were fitted with stomas in the abdominal wall, connected to the terminal end of the small intestine. The intestinal effluent was excreted through the stoma into a collection bag, enabling periodic collection of digesta for analysis of starch, energy, nonstarch polysaccharides, and other nutrient components. Only 2% of starch remained undigested after finely milled barley was eaten, but 17% of flaked barley resisted digestion, mostly as intact starch granules. Energy excretion from the stoma was three times higher after flaked than after milled barley, indicating differences in glycemic responses due to physical form and/or processing effects. Björck et al. (1994) reviewed food properties affecting digestion and absorption of carbohydrates, pointing out that any process that disrupts the physical or botanical structure of food ingredients will increase glycemic response levels. This includes the addition of ingredients with viscosity properties, indigestible characteristics such as RS or complexes (Holm et al. 1983). Various degrees of heat and moisture treatments of oat and barley flakes were also evaluated by this group (Granfeldt et al. 2000). Tosh et al. (2007) reported an in vitro method to predict glycemic responses of foods using viscosity as a function of β-glucan concentration and molecular weight. This method has the potential to estimate glycemic response to barley foods without costly and time-consuming clinical trials.

Second-Meal Effect

The studies described above have addressed the desirability of achieving a lowered blood glucose peak postprandially, or immediately after a meal. Considerable interest has been shown in improving insulin economy in the meal following consumption of a food, termed the *second-meal effect* (Jenkins et al. 1982; Wolever et al. 1988; Brighenti et al. 2006). Two mechanisms are associated with long-term metabolic effects: (1) a decreased demand for insulin due to slow digestive and absorptive phases (lente carbohydrate), and (2) formation of SCFAs from fermentation of dietary fiber (DF) or RS in the colon. The SCFAs, especially propionate, are believed to affect glucose metabolism in a beneficial manner (Anderson and Bridges 1984). Liljeberg et al. (1999) served breakfast meals containing bread made from high-amylose Glacier barley and flaked PW barley to healthy subjects, using white wheat bread as a reference. Barley breads contained 70% high-amylose whole-meal barley flour and 30% white wheat flour. The test meals included butter, cheese, and milk; all contained 50 g of available starch, and the two test meals had GI values of 71 and 60. Glucose and insulin responses were measured following the breakfast meals and satiety scores for the breakfast meals were estimated. Four hours after the breakfast, the subjects were served a standard high-GI lunch, and the glucose and insulin responses were remeasured. The two earlier test meals with low GI values improved glucose tolerance at the second meal. The highest satiety score was found with the barley breakfast, with low GI and high RS and DF content. In this particular study, the improved glucose tolerance at the second meal was attributed to slow absorption and digestion of starch from the breakfast meal. However, other reports in the literature at that time were associated with the carbohydrate fermentation concept, whereby SCFA production suppresses hepatic glucose production (Wolever et al. 1995; Thorburn et al. 1993). Östman et al. (2002) reported that barley bread containing lactic acid improved glucose tolerance at a subsequent meal. A similar improvement in glucose response was observed when fermented sourdough bread was fed (Liljeberg et al. 1995). The relative importance of acid addition to bread by either fermentation or addition as an ingredient to the glucose response has not been elucidated and needs further assessment as a practical concern in food product development.

The University of Lund researchers subsequently reported some methodology improvements of the second-meal technique applied to the effect of a low-GI evening meal on glycemic response to the following morning breakfast meal (Granfeldt et al. 2006). This study calculated available carbohydrate load as total carbohydrates minus the indigestible components DF and RS. The evening meals varied in GI and/or indigestible carbohydrates, and consisted of pasta, barley kernels, and white wheat bread. Serving sizes were adjusted to correspond with the DF and RS content. The test evening meals were found to substantially reduce the GI and insulinemic index of white wheat bread eaten at the subsequent breakfast meal. Following this study, Nilsson et al. (2007) expanded the study design to include four different low-GI evening meals and a white bread control. Tests following the subsequent breakfast meal were expanded to include measurement of free fatty acid levels, and the markers of colonic fermentation SCFAs in blood and urine and expired H_2. The after-breakfast glucose response was significantly lower following consumption of barley kernels the previous evening than after white bread or spaghetti with wheat bran. Breath H_2 excretion and plasma SCFAs were higher at breakfast when subjects had previously consumed barley, and FFAs were lower, in accordance with earlier reports. These results indicate that colonic fermentation of the specific indigestible carbohydrates present as well as low-GI features may both influence the improved glucose tolerance at breakfast. A follow-up study (Nilsson et al. 2007) was conducted to elucidate the mechanism of improved glucose tolerance at breakfast. Test evening meals consisted of high-GI white bread (WWB), WWB + barley + barley DF corresponding to the DF of barley kernels, low-GI spaghetti + barley DF, spaghetti + double amounts of barley DF (2*DF), spaghetti + oat DF, or whole-grain barley flour porridge. The rationale of the treatment containing spaghetti + double the barley DF (2*DF) was to mimic the DF content of the whole barley kernels in the earlier studies. At breakfast the morning after the test meals, blood glucose, insulin, SCFAs, FFAs and breath H_2 were measured. The most effective evening meal for lowering glucose response was spaghetti + 2*DF. This meal also resulted in the highest breath H_2 excretion. The glucose incremental AUC after breakfast was correlated positively with fasting FFAs, suggesting colonic fermentation as a modulator of glucose tolerance through a mechanism leading to suppressed FFA levels. This study demonstrated that the low GI of a food per se is not attributable to slow digestion and absorption of sufficient magnitude to induce overnight benefits. These researchers suggested the possibility that the presence of abundant fermentable fiber in a food may be a more reliable indicator of potential low glycemic responses. More recently (Nilsson et al. 2008), this group repeated the study using both boiled barley kernels and WWB + barley fiber + RS at an evening meal. At breakfast, glucose response was negatively correlated with colonic fermentation and GLP-1 and positively correlated with fasting FFA. The levels of the pro-inflammatory marker IL-6 were lower at breakfast, and fasting adiponectin was higher after the previous barley kernel evening meals. These results support their hypothesis of high fermentable fiber as the definitive factor in physiological responses in carbohydrate metabolism.

BENEFICIAL EFFECTS OF RESISTANT STARCH IN BARLEY ON THE INTESTINAL TRACT

The indigestible fiber components of barley, especially β-glucan, as well as RS, progress through the digestive tract on into the large intestine. Fermentation of this material by microflora then occurs, resulting in formation of short-chain fatty acids (SCFA), especially butyrate and propionate (Topping and Clifton 2001).The benefits of these fatty acids in the large intestine are healthy colonic mucosa, and provision of an energy source for epithelial cells (Topping et al. 2003). Dongowski et al. (2002) investigated the effects of a high-amylose barley and PW barley, compared with maize starch, a high-RS commercial product, and a control diet with no barley. These materials were extruded to prepare experimental diets for rats. All of the animals fed barley diets thrived better and had greater intestinal mass than controls. SCFAs (i.e., acetic, propionic, and butyric) were higher in cecal and colon contents of animals fed the test diets. The more acid conditions indicated a smaller proportion of secondary bile acids, believed to be promoting factors in colon cancer. Bird et al. (2004b) fed stabilized whole-grain barley flours from Himalaya 292, the high-amylose barley cited previously (Bird et al. 2004a), with two other barleys, and either wheat or oat bran to rats, to determine the effects of these feeds on intestinal SCFAs. Although there were independent differences between animals fed different diets, colonic SCFAs were consistently higher and pH was lower in those fed Himalaya 292 barley than all other diets. Results were attributed to the greater RS content in this barley cultivar, due to its reduced amylopectin and increased amylose content as well as having a high β-glucan content. More recently, Bird et al. (2007) compared the effects of foods made from whole-grain Himalaya 292 with those made from whole-grain wheat or refined cereal. Biomarkers of bowel health in healthy subjects (fecal weight, fecal concentration of butyrate, SCFA excretion, and fecal p-cresol concentration) were all significantly different between groups and indicative of improved bowel health in subjects who consumed barley diets. The beneficial health benefits were attributed to the presence of resistant starch.

SUMMARY

Barley has been shown in animal studies and human clinical trials to be an ordinary food product that can effectively normalize blood cholesterol. β-Glucan, a significant part of the dietary fiber in barley, is the most important component for lowering blood cholesterol concentration. It is generally believed that this beneficial effect is the result of increased intestinal viscosity, delaying absorption of food fat and binding bile acids, which cycle with body cholesterol. Data from studies with extracted β-glucans have not been as consistent in cholesterol-lowering efficacy as with β-glucans in the natural state in the grain or grain products. Further research in extraction procedures to obtain semipurified β-glucan with intact physiochemical properties will no doubt improve the efficacy of the products.

Diabetes and metabolic syndrome are contemporary challenges in public health. It is well known that wise, selective choices and appropriate quantities of carbohydrate foods are dramatically effective in controlling glucose and insulin responses to food intake. Consumption of barley foods have been shown in clinical trials to lower blood glucose curves, even over several hours, and to influence glucose metabolism at the following meal. The same soluble fiber components in barley that are effective in lowering blood cholesterol, β-glucan, and arabinoxylans also affect glycemic responses. In addition, varieties of barley having a high amylose/amylopectin ratio tend to form resistant starch during processing. Both soluble fiber and resistant starch are indigestible carbohydrates which undergo fermentation in the large intestine, producing the short-chain fatty acids, particularly butyrate, which enhance cell integrity in the bowel. Barley's reputation in ancient history appears to be well founded, in view of its health benefits in three important physiological and metabolic areas: lipid metabolism, glycemic control, and intestinal health.

REFERENCES

Ajgaonker, S. S. 1972. Diabetes mellitus as seen in the ancient Ayurvedic medicine. Pages 13–20 in: *Insulin and Metabolism*. J. S. Bajaj, ed. Association of India, Bombay, India.

AHA (American Heart Association). 1988. Dietary guideline for healthy adults: a statement for physicians and health professionals by the Nutrition Committee, AHA. *Circulation* 77:721A–724A.

Åkerberg, A., Liljeberg, H., and Björck, I. 1998. Effects of amylose/amylopectin ratio and baking conditions on resistant starch formation and glycaemic indices. *J. Cereal Sci.* 20:71–80.

Anderson, J. W., and Bridges, S. R. 1984. Short-chain fatty acid fermentation products of plant fiber affect glucose metabolism of isolated rat hepatocytes. *Proc. Soc. Exp. Biol. Med.* 177:372–376.

Araki, S., Aoe, S., Kato, M., Kihara, M., Shimizu, C., Nakamura, Y, Ito, K., Hayashi, K., Watari, J., and Ikegami, S. 2007. Effect of β-glucan-enriched barley on health benefits: II. Studies of cholesterol metabolism and visceral fat in STR/Ort mice. Page 110 in: *Proc. Annual Meeting of the Japan Society for Bioscience, Biotechnology, and Agrochemistry*, Tokyo, Mar. 25–27.

Asp, N.-G. 1992. Resistant starch. Proceedings from the second plenary meeting of EURESTA: European FLAIR. Concerted Action No. 11 on physiological implications of the consumption of resistant starch in man (preface). *Eur. J. Clin. Nutr.* 46(Suppl. 2):S1.

Battilana, P., Omstan, K., Minehira, K., Schwarz, J. M., Acheson, K., Schneiter, P., Burri, J., Jeqular, E., and Tappy, L. 2001. Mechanisms of action of β-glucan in postprandial glucose metabolism in healthy men. *Eur. J. Clin. Nutr.* 56:327–333.

Behall, K. M., and Hallfrisch, J. 2006. Effects of barley consumption on CVD risk factors. *Cereal Foods World* 51:12–15.

Behall, K. M., Scholfield, D. J., Yuhaniak, I., and Canary, J. 1989. Diets containing high amylose vs. amylopectin starch: effects on metabolic variables in human subjects. *Am. J. Clin. Nutr.* 49:337–344.

Behall, K. M., Scholfield, D. J., and Hallfrisch, J. 2004a. Lipids significantly reduced by diets containing barley in moderately hypercholesterolemic men. *J. Am. Coll. Nutr.* 23:55–62.

Behall, K. M., Scholfield, D. J., and Hallfrisch, J. 2004b. Diets containing barley significantly reduce lipids in mildly hypercholesterolemic men and women. *Am. J. Clin. Nutr.* 80:1185–1193.

Behall, K. M., Scholfield, D. J., and Hallfrisch, J. 2005. Comparison of hormone and glucose responses of overweight women to barley and oats. *J. Am. Coll. Nutr.* 24:182–188.

Behall, K. M., Scholfield, D. J., Hallfrish, J. G., and Liljeberg-Elmstahl, H. 2006a. Consumption of both resistant starch and β-glucan improves postprandial plasma glucose and insulin in women. *Diabetes Care* 29:976–981.

Behall, K. M., Scholfield, D. J., and Hallfrisch, J. G. 2006b. Barley β-glucan reduces plasma glucose and insulin responses compared with resistant starch in men. *Nutr. Res.* 26:644–650.

Bengtsson, S., Åman, P., Graham, H., Newman, C. W., and Newman, R. K. 1990. Chemical studies on mixed-linked β-glucans in hull-less barley cultivars giving different hypercholesterolaemic responses in chickens. *J. Sci. Food Agric.* 52:435–445.

Bird, A. R., Jackson, M., King, R. A., Davies, D. A., Usher, S., and Topping, D. L. 2004a. A novel high-amylose barley cultivar (*Hordeum vulgare* var.) Himalaya 292 lowers plasma cholesterol and alters indices of large-bowel fermentation in pigs. *Brit. J. Nutr.* 92:607–615.

Bird, A. R., Flory, C., Davies, D. A., Usher, S., and Topping, D. L. 2004b. A novel barley cultivar (Himalaya 292) with a specific gene mutation in starch synthase IIa raises large bowel starch and short-chain fatty acids in rats. *J. Nutr.* 134:831–835.

Bird, A. R., Vuaran, M. S., King, R. A., Noakes, M., Keogh, J., Morell, M. K., and Topping, D. L. 2007. Wholegrain foods from a high-amylose barley variety (Himalaya 292) improve indices of bowel health in human subjects. *Brt. J. Nutr. Online Publ.*, Oct. 8.

Björck, I., and Elmståhl, H. L. 2003. The glycaemic index: importance of dietary fibre and other food properties. *Proc. Nutr. Soc.* 62:201–206.

Björck, I., Eliasson, A.-C., Drews, A., Gudmundsson, M., and Karlsson, R. 1990. Some nutritional properties of starch and dietary fiber in barley genotypes containing different levels of amylose. *Cereal Chem.* 67:327–333.

Björck, I., Granfeldt, Y., Liljeberg, H., Tovar, J., and Asp N.-G. 1994. Food properties affecting the digestion and absorption of carbohydrates. *Am. J. Clin. Nutr.* 59(Suppl.): 699S–705S.

Björklund, M., van Rees, A., Mensink, R. P., and Önning, G. 2005. Changes in serum lipids and postprandial glucose and insulin concentrations after consumption of beverages with β-glucans from oats or barley: a randomized dose-controlled trial. *Eur. J. Clin. Nutr.* 59:1272–1281.

Botham, R. L., Cairns, P., Faulks, R. M., Livesey, G., Morris, V. J., Noel, T. R., and Ring, S. G. 1997. Physicochemical characterization of barley carbohydrates resistant to digestion in a human ileostomate. *Cereal Chem.* 74:29–33.

Bourdon, I., Yokayama, W., Davis, P., Hudson, C., Backus, R., Richter, D., Knuckles, B., and Schneeman, B. O. 1999. Postprandial lipid, glucose, insulin and cholecystokinin responses in men fed barley pasta enriched with β-glucan. *Am. J. Clin. Nutr*. 69:55–63.

Brand-Miller, J. C. 1994. Importance of glycemic index in diabetes. *Am. J. Clin. Nutr*. 59(Suppl.): 747S–752S.

Brand-Miller, J. 2007. The glycemic index as a measure of health and nutritional quality: an Australian perspective. *Cereal Foods World* 52:41–44.

Brennan, C. S., and Cleary, L. J. 2005. The potential use of cereal (1 → 3, 1 → 4)-β-D-glucans as functional food ingredients. *J. Cereal Sci*. 42:1–13.

Brighenti, F., Benini, L., Del Rio, D., Casiraghi, C., Pellegrini, N., Scazzina, F., Jenkins, D. J. A., and Vantini, I. 2006. Colonic fermentation of indigestible carbohydrates contributes to the second-meal effect. *Am. J. Clin. Nutr*. 83:817–822.

Cavallero, A., Empilli, S., Brighenti, F., and Stanca, A. M. 2002. High (1 → 3, 1 → 4)-β-glucan barley fractions in breadmaking and their effects on human glycemic response. *J. Cereal Sci*. 36:59–66.

CDC. (U.S. Centers for Disease Control and Prevention). 2007. http://www.cdc.gov/DataStatistics.

Chaudhary, V. K., and Weber, F. E. 1990. Barley bran flour evaluated as dietary fiber ingredient in wheat bread. *Cereal Foods World* 35:560–562.

Cleary, L., and Brennan, C. 2006. The influence of a (1 → 3)(1 → 4)-β-D-glucan rich fraction from barley on the physico-chemical properties and in vitro reducing sugars release of durum wheat pasta. *Int. J. Food Sci. Technol*. 41:910–918.

Danielson, A. D., Newman, R. K., Newman, C. W., and Berardinelli, J. G. 1997. Lipid levels and digesta viscosity of rats fed a high-fiber barley milling fraction. *Nutr. Res*. 17:515–522.

Darwiche, G., Ostman, E. M., Liljeberg, H. G., Kalliner, N., Björgell, O., Björck, I. M., and Almer, L. O. 2001. Measurements of gastric emptying rate by use of ultrasonography: studies in humans using bread with added sodium propionate. *Am. J. Clin. Nutr*. 74:254–256.

DeGroot, A. P., Luyken, R., and Pikaar, N. A. 1963. Cholesterol-lowering effect of rolled oats. *Lancet* 2:303–304.

Delaney, B., Nicolosi, R. J., Wilson, T. A., Carlson, T., Frazer, S., Zheng, G.-H., Hess, R., Ostergren, K., Haworth, J., and Knutson, N. 2003. β-Glucan fractions from barley and oats are similarly antiatherogenic in hypercholesterolemic Syrian golden hamsters. *J. Nutr*. 133:468–475.

DeVries, J. W. 2007. Glycemic index: the analytical perspective. *Cereal Foods World* 52:45–49.

Dongowski, G., Huth, M., Gebhardt, E., and Flamme, W. 2002. Dietary fiber-rich barley products beneficially affect the intestinal tract of rats. *J. Nutr*. 132:3704–3714.

Fadel, J., Newman, R. K., Newman, C. W., and Barnes, A. J. 1987. Hypocholesterolemic effects of beta-glucans in different barley diets fed to broiler chicks. *Nutr. Rep. Int*. 35:1049–1058.

Ferrannini, E., Barrett, E. J., Bevilacqua, S., and DeFronzo, R. A. 1983. Effect of fatty acids on glucose production and utilization in man. *J. Clin. Invest*. 72:1737–1747.

Fisher, H., and Griminger, P. 1967. Cholesterol lowering effects of certain grains and of oat fractions in chickens. *Proc. Soc. Exp. Biol. Med*. 126:108–111.

Freedman, D. S., Otvos, J. D., Jeyarajah, E. J., Darboriak, J. J., Anderson, A. J., and Walker, J. A. 1998. Relation of lipoprotein subclasses as measured by protein nuclear magnetic resonance spectroscopy to coronary artery disease. *Arterioscler. Thromb. Vasc. Biol.* 18:1046–1053.

Garwiche, G., Östman, E. M., Liljeberg, H. G., Kallinen, N., Björgell, O., Björck, I., and Almer, L. O. 2001. Measurements of the gastric emptying rate by use of ultrasonography: studies in humans using bread with added sodium propionate. *Am. J. Clin. Nutr.* 74:254–258.

Gentilcore, D., Chaikomen, R., Jones, K. L., Russo, A., Feinle-Bisset, C., Wishart, J. M., Rayner, C. K., and Horowitz, M. 2006. Effects of fat on gastric emptying of and the glycemic, insulin and incretin responses to a carbohydrate meal in type 2 diabetes. *J. Clin. Endocrinol. Metab.* 91:2062–2067.

Goddard, M. S., Young, G., and Marcus, R. 1984. The effect of amylose content on insulin and glucose response to ingested rice. *Am. J. Clin. Nutr.* 39:388–392.

Granfeldt, Y., Björck, I., Drews, A., and Tovar, J. 1992. An in vitro procedure based on chewing to predict metabolic response to starch in cereal and legume products. *Eur. J. Clin. Nutr.* 46:649–660.

Granfeldt, Y., Liljeberg, H., Drews, A., Newman, R., and Björck, I. 1994. Glucose and insulin responses to barley products: influence of food structure and amylose–amylopectin ratio. *Am. J. Clin. Nutr.* 59:1075–1082.

Granfeldt, Y., Eliasson, A.-C., and Björck, I. 2000. An examination of the possibility of lowering the glycemic index of oat and barley flakes by minimal processing. *J. Nutr.* 130:2207–2214.

Granfeldt, Y., Wu, X., and Björck, I. 2006. Determination of glycaemic index; some methodological aspects related to the analysis of carbohydrate load and characteristics of the previous evening meal. *Eur. J. Clin. Nutr.* 60:104–112.

Grundy, S. M. and Denke, M. A. 1990. Dietary influences on serum lipids and lipoproteins. *Lipid Res.* 31:1149–1172.

Hallfrisch, J., Scholfield, D. J., and Behall, K. M. 2003. Physiological responses of men and women to barley and oat extracts (Nu-trimX): II. Comparison of glucose and insulin responses. *Cereal Chem.* 80:80–83.

Hegsted, D. M., McGandy, R. B., Myers, M. K., and Stare, F. J. 1965. Quantitative effects of dietary fat on serum cholesterol in man. *Am. J. Clin. Nutr.* 17:281–295.

Hinata, M., Ono, M., Midoikawa, S., and Nakanishi, K. 2007. Metabolic improvement of male prisoners with type 2 diabetes in Fukushima Prison, Japan. *Diabetes Res. Clin. Pract.* 77:327–332.

Holm, J., Björck, I., Ostrowska, S., Eliasson, A.-C., Asp, N.-G., Larsson, K., and Lundquist, I. 1983. Digestibility of amylose–lipid complexes in vitro and in vivo. *Starch/Stärke* 35:294–297.

Horowitz, M., Edelbroek, M. S., Wishart, J. M. and Straatof, J. W. 1993. Relationship between oral glucose tolerance and gastric emptying in normal healthy subjects. *Diabetologia* 36:857–862.

Hullinger, C. H., Van Patten, E., and Freck, J. A. 1973. Food applications of high amylose starches. *Food Technol.* 27:22–28.

Huth, M., Dongowski, G., Gebhardt, E., and Flamme, W. 2000. Functional properties of dietary fibre enriched extrudates from barley. *J. Cereal Sci.* 32:115–128.

Ikegami, S., Tsuchihashi, F., Nakamura, K., and Innami, S. 1991. Effect of barley on development of experimental diabetes in rats. *J. Jpn. Soc. Nutr. Food Sci*. 44:447–454.

Ikegami, S., Tomita, M., Honda, S., Yamaguchi, M., Mizukawa, R., Suzuki, Y., Ishii, K., Ohsawa, S., Kiyooka, N., Higuchi, M., and Kobayashi, S. 1996. Effect of boiled barley-rice-feeding in hypercholesterolemic and normelipemic subjects. *Plant Foods Hum. Nutr*. 49:317–328.

Inglett, G. E. 2000. Soluble hydrocolloid food additives and method of making. U.S. patent 6,060,519.

Jenkins, A. L. 2007. The glycemic index:looking back 25 years. *Cereal Foods World* 52:53.

Jenkins, D. J. A., Wolever, T. M. S., Taylor, R. H., Griffiths, C., Krzeminska, K., Lawrie, J. A., Bennett, C. M., Goff, D. V., Sarson, D. L., and Bloom, S. R. 1982. Slow release dietary carbohydrate improves second meal tolerance. *Am. J. Clin. Nutr*. 35:1339–1346.

Jenkins, D. J. A., Wolever, T. M. S., Venketeshwer Rao, A., Hegele, R. E., Mitchell, S. J., Ransom, T. P. P., Boctor, D. L., Spadafora, P. J., Jenkins, A. L., Mehling, C., Relle, L. K., Connelly, P. W., Story, J. A., Furumoto, E. J., Corey, P., and Würsch, P. 1993. Effect on blood lipids of very high intakes of fiber in diets low in saturated fat and cholesterol. *N. Engl. J. Med*. 329:21–26.

Jenkins, D. J. A., Wolever, T. M. S., Leeds, A. R., Gassull, M. A., Dilawn, J. B., Hansman, P., Dilawan, J., Goff, D. V., Metz, G. L., and Alberti, K. G. M. M. 1998. Dietary fibers, fiber analogues and glucose tolerance: importance of viscosity. *Brt. Med. J*. 1:1392–1394.

Jenkins, D. J. A., Kendall, C. W. C., Marchie, A., Faulkner, D. A., Wong, J. M. W., deSouza, R., Emam, A., Parker, T. L., Vidgen, E., Trautwein, E. A., Lapsley, K. G., Josse, R. G., Leiter, L. A., Singer, W., and Connelly, P. W. 2005. Direct comparison of a dietary portfolio of cholesterol-lowering foods with a statin in hypercholesterolemic participants. *Am. J. Clin. Nutr*. 81:380–387.

Jones, J. M. 2007. The AACC International glycemic response definitions. *Cereal Foods World* 52:54–55.

Kahlon, T. S., and Woodruff, C. L. 2003. In vitro binding of bile acids by rice bran, oat bran, barley and beta-glucan enriched barley. *Cereal Chem*. 80:260–263.

Kahlon, T. S., Chow, F. L., Knuckles, B. E., and Chiu, M. M. 1993. Cholesterol-lowering effects in hamsters of β-glucan-enriched barley fractions, dehulled whole barley, rice bran, and oat bran and their combinations. *Cereal Chem*. 70:435–439.

Kalra, S., and Jood, S. 2000. Effect of barley β-glucan on cholesterol and lipoprotein fractions in rats. *J. Cereal Sci*. 31:141–145.

Keenan, J. M., Goulson, M., Shamliyan, T., Knutson, N., Kolberg, L., and Curry, L. 2007. The effects of concentrated barley β-glucan on blood lipids in a population of hypocholesterolemic men and women. *Brt. J. Nutr*. 97:1162–1168.

Keogh, G. F., Cooper, G. J. S., Mulvey, T. B., McArdle, B. H., Coles, G. D., Monro, J. A., and Poppitt, S. D. 2003. Randomized controlled crossover study of the effect of a highly β-glucan-enriched barley on cardiovascular disease risk factors in mildly hypercholesterolemic men. *Am. J. Clin. Nutr*. 78:711–718.

Keogh, J. B., Lau, C. W. H., Noakes, M., Bowen, J., and Clifton, P. M. 2007. Effects of meals with high soluble fibre, high amylose barley variant on glucose, insulin, satiety and thermic effect of food in healthy lean women. *Eur. J. Clin. Nutr.* 61:597–604.

Kerckhoffs, D. A. J. M., Hornstra, G., and Mensink, R. P. 2003. Cholesterol-lowering effect of β-glucan from oat bran in mildly hypercholesterolemic subjects may decrease when β-glucan is incorporated into bread and cookies. *Am. J. Clin. Nutr.* 78:221–227.

Keys, A. 1970. Coronary heart disease in seven countries. *Circulation* 41:I-1 and I-2.

Keys, A., Anderson, J. T., and Grande, F. 1957. Prediction of serum-cholesterol responses of man to changes in fats in the diet. *Lancet* 2:959–961.

Kim, H., Behall, K. M., Vinyard, B., and Conway, J. M. 2006. Short-term satiety and glycemic response after consumption of whole grains with various amounts of β-glucan. *Cereal Foods World* 51:29–33.

Kim, J.-Y., van de Wall, E., Laplante, M., Azzara, A., Trujillo, M. E., Hofmann, S. M., Schraw, T., Durand, J. L., Li, H., Li, G., Jelicks, L. A., Mehler, M. F., Hui, D. Y., Deshaies, Y., Shulman, G. I., Schwartz, G. J., and Scherer, P. E. 2007. Obesity-associated improvements in metabolic profile through expansion of adipose tissue. *J. Clin. Invest.* 117:2621–2637.

Kishimoto, T. 1989. The biology of interleukin-6. *Blood* 74:1–10.

Knuckles, B. E., Hudson, C. A., and Chiu, M. M. 1997. Effect of β-glucan barley fractions in high-fiber bread and pasta. *Cereal Foods World* 42:94–99.

Li, J., Kaneko, T., Qin, L.-Q., Wang, J., Wang, Y., and Sato, A. 2003a. Long-term effects of high dietary fiber intake on glucose tolerance and lipid metabolism in GK rats: comparison among barley, rice and cornstarch. *Metabolism* 62:1206–1210.

Li, J., Kaneko, T., Qin, L.-Q., Wang, J., and Wang, Y. 2003b. Effects of barley intake on glucose tolerance, lipid metabolism and bowel function in women. *Nutrition* 19: 926–929.

Lia, A., Hallmans, G., Sandberg, A. S., Slundberg, B., Åman, P., and Andersson, H. K. 1995. Oat beta-glucan increases bile acid excretion and a fiber-rich barley fraction increases cholesterol excretion in ileostomy subjects. *Am. J. Clin. Nutr.* 62:245–1251.

Lifschitz, C. H., Grusak, M. A., and Butte, N. F. 2002. Carbohydrate digestion in humans from a β-glucan-enriched barley is reduced. *J. Nutr.* 132:2593–2596.

Liljeberg, H., and Björck, I. 1994. Bioavailability of starch in bread products: postprandial glucose and insulin responses in healthy subjects and in vitro resistant starch content. *Eur. J. Clin. Nutr.* 48:151–163.

Liljeberg, H. G., Lonner, C. H. and Björck, I. M. 1995. Sourdough fermentation or addition of organic acids or corresponding salts to bread improves nutritional properties of starch in healthy humans. *J. Nutr* 125:1503–1511.

Liljeberg, H. F. M., Granfeldt, Y. E., and Björck, I. M. E. 1996. Products based on high fiber barley genotype, but not on common barley or oats, lower postprandial glucose and insulin responses in healthy humans. *J. Nutr.* 126:458–466.

Liljeberg, H. G. M., Åkerberg, A. K. E., and Björck, I. M. E. 1999. Effect of the glycemic index and content of indigestible carbohydrates of cereal-based breakfast meals on glucose tolerance at lunch in healthy subjects. *Am. J. Clin. Nutr.* 69:647–655.

Livesey, G., Wilkinson, J. A., Roe, M., Faulks, R., Clark, S., Brown, J. C., Kennedy, H., and Elia, M. 1995. Influence of the physical form of barley grain on the digestion of its starch in the human small intestine and implications for health. *Am. J. Clin. Nutr.* 61:75–81.

Lupton, J. R., Robinson, M. C., and Morin, J. L. 1994. Cholesterol-lowering effect of barley bran flour and oil. *J. Am. Diet. Assoc.* 94:65–70.

Macdonald, I. 1965. The effect of various dietary carbohydrates on the serum lipids during a five day regimen. *Clin. Sci.* 29:193–197.

Martinez, V. M., Newman, R. K., and Newman, C. W. 1992. Barley diets with different fat sources have hypocholesterolemic effects in chicks. *J. Nutr.* 122:1070–1076.

McIntosh, G. H., Whyte, J., McArthur, R., and Nestel, P. J. 1991. Barley and wheat foods: influence on plasma cholesterol concentration in hypercholesterolemic men. *Am. J. Clin. Nutr.* 53:1205–1209.

McIntosh, G. H., Newman, R. K., and Newman, C. W. 1995. Barley foods and their influence on cholesterol metabolism. Pages 89–108 in: *Plants in Human Nutrition.* World Review of Nutrition and Dieteties 77. A. P. Simopoulus, ed. S. Karger, Basel, Switzerland.

Milton, J. E., Sananthanan, C. S., Patterson, M., Ghatei, M. S., Bloom, S. R., and Frost, G. S. 2007. Glucagon-like peptide-1 (7–36) amide response to low versus high glycaemic index preloads in overweight subjects with and without type II diabetes mellitus. *Eur. J. Clin. Nutr.* Published online Feb. 14.

Mori, T. 1990. Chemical characterization and metabolic function of soluble dietary fiber from select milling fractions of a hull-less barley and its waxy starch mutant. M.S. thesis. Montana State University, Bozemon, MT.

Narain, J. P., Shukla, R. L., Biflani, K. P., Kochnar, K. P., Karmarkar, M. G., Bala, S., Srivastara, L. M., and Reddy, K. S. 1992. Metabolic responses to a four week barley supplement. *Int. J. Food Sci. Nutr.* 43:41–46.

Nauck, M. A., Baller, B., and Meier, J. J. 2004. Gastric inhibitory polypeptides and glucagon-like peptide-1 in the pathogenesis of type 2 diabetes. *Diabetes* 53(Suppl. 3): S190–S196.

Naumann, E., van Rees, A. B., Önning, G., Öste, R., Wydra, M., and Mensink, R. P. 2006. β-Glucan incorporated into a fruit drink effectively lowers serum LDL-cholesterol concentrations. *Am. J. Clin. Nutr.* 83:601–605.

NCEP [National Cholesterol Education Program Expert Panel on Detection, Evaluation and Treatment of High Blood Cholesterol in Adults (Adult Treatment Panel III)]. 2001. *J. Am. Med. Assoc.* 285:2486–2497.

Newman, R. K., Lewis, S. E., Newman, C. W., Boik, R. J., and Ramage, R. T. 1989a. Hypocholesterolemic effect of barley food and healthy men. *Nutr. Rep. Int.* 39:749–760.

Newman, R. K., Newman, C. W., and Graham, H. 1989b. The hypocholesterolemic function of barley β-glucans. *Cereal Foods World* 34:883–886.

Nilsson, A., Granfeldt, Y., Östman, E., Preston, T., and Björck, I. 2007. Effects of GI and content of indigestible carbohydrates of cereal-based evening meals on glucose tolerance at a subsequent standardized breakfast. *Eur. J. Clin. Nutr.* Published online May 23.

Nilsson, A., Östman, E. M., Holst, J. J. and Björck, I. M. 2008. Including indigestible carbohydrates in the evening meal of healthy subjects improves glucose tolerance, lowers inflammatory markers, and increases satiety after a subsequent standardized breakfast. *J. Nutr.* 138:732–739.

Östman, E. M., Liljeberg, H. G. M., and Björck, I. M. E. 2002. Barley bread containing lactic acid improves glucose tolerance at a subsequent meal in healthy men and women. *J. Nutr.* 132:1173–1175.

Östman, E., Rossi, E., Larsson, H., Brighenti, F., and Björck, I. 2006. Glucose and insulin responses in healthy man to barley bread with different levels of (1 → 3; 1 → 4)-β-glucans: predictions using fluidity measurement of in vitro enzyme digests. *J. Cereal Sci.* 43:230–235.

Pick, M. E., Hawrysh, Z. J., Gee, M. I., and Toth, E. 1998. Barley bread products improve glycemic control of type 2 subjects. *Int. J. Food Sci. Nutr.* 49:71–81.

Pins, J. J., and Kaur, H. 2006. A review of the effects of barley β-glucan on cardiovascular and diabetic risk. *Cereal Foods World* 51:8–11.

Poppitt, S. D. 2007. Soluble oat and barley β-glucan enriched products: Can we predict cholesterol-lowering effects? *Brt. J. Nutr.* 97:1049–1050.

Poppitt, S. D., van Drunen, J. D. E., McGill, A.-T., Mulvey, T. B., and Leahy, F. E. 2007. Supplementation of a high-carbohydrate breakfast with barley β-glucan improves postprandial glycaemic response for meals but not beverages. *Asia Pac. J. Clin. Nutr.* 16:16–24.

Rendell, M., Vanderhoof, J., Venn, M., Shehas, M. A., Arndt, E., Rao, C. S., Gill, G., Newman, R. K., and Newman, C. W. 2005. Effect of a barley breakfast cereal on blood glucose and insulin response in normal and diabetic patients. *Plant Foods Hum. Nutr.* 60:63–67.

Rimm, E. B., Ascherio, A., Giovannucci, F., Spiegelman, D., Stampfer, M. J., and Willett, W. 1996. Vegetable, fruit and cereal fiber intake and risk of coronary heart disease among men. *J. Am. Med. Assoc.* 275:447–451.

Salmeron, J., Manson, J. E., Stampler, M. J., Colditz, G. S., Wing, A. L., and Willett, W. O. 1997. Dietary fiber, glycemic load, and risk of non-insulin-dependent diabetes mellitus in women. *J. Am. Med. Assoc.* 277:472–477.

Sane, T., and Nikkila, E. A. 1988. Very low density lipoprotein triglyceride metabolism in relatives of hypertriglyceridemia probands. *Arteriosclerosis* 8:217–226.

Sato, J., Osawa, I., Hattori, Y., Oshida, Y., and Sato, Y. 1990. Effects of dietary fiber on carbohydrate metabolism—a study in healthy subjects and diabetic patients. *Nagoya J. Health, Physical Fitness Sports* 13:75–78.

Shimizu, C., Kihara, M., Aoe, S., Araki, S., Ito, K., Hayashi, K, Watari, J., Sakata, Y., and Ikegama, S. 2007. Effect of high β-glucan barley on serum cholesterol concentrations and visceral fat area in Japanese men: a randomized, double-blinded, placebo-controlled trial. *Plant Foods Hum. Nutr.* 63:21–25.

Shukla, K., Narain, J. P., Puri, P., Gupta, A., Bijlani, R. L., Mahapatra, S. C., and Karmarkar, M. G. 1991. Glycemic response to maize, bajra and barley. *Indian J. Physiol. Pharmacol.* 35:249–254.

Thorburn, A., Muir, J., and Proietto, J. 1993. Carbohydrate fermentation decreases hepatic glucose output in healthy subjects. *Metabolism* 42:780–785.

Tietyen, J. L., Nevins, D. J., and Schneeman, B. O. 1990. Characterization of the hyper-cholesterolemic potential of oat bran. *FASEB J.* 4:A527.

Topping, D. L., and Clifton, P. M. 2001. Short-chain fatty acids and human colonic function: roles of resistant starch and nonstarch polysaccharides. *Physiol. Rev.* 81:1031–1054.

Topping, D. L., Morell, M. K., King, R. A., Zhongyi, L., Bird, A. R., and Noakes, M. 2003. Resistant starch and health: Himalaya 292, a novel barley cultivar to deliver benefits to consumers. *Starch/Stärke* 55:539–545.

Tosh, S. M., Wood, P. J., and Wolever, T. M. 2007. Use of in vitro extraction of β-glucan to predict glycemic response (abstr.). *Cereal Foods World* 52:A30.

Trowell, H. 1973. Dietary fibre, ischaemic heart disease and diabetes mellitus. *Proc. Nutr. Soc.* 32:151–157.

Trowell, H. 1975. Coronary heart disease and dietary fiber. *Am. J. Clin. Nutr.* 28:798–800.

Truswell, A. S. 2002. Cereal grains and coronary heart disease. *Eur. J. Clin. Nutr.* 56:1–14.

Wang, L., Newman, R. K., Newman, C. W., and Hofer, P. J. 1992. Barley β-glucan alters intestinal viscosity and reduces plasma cholesterol concentration in chicks. *J. Nutr.* 122:2292–2297.

Wolever, T. M. S. 1990. The glycemic index. *World Rev. Nutr. Diet.* 62:120–185.

Wolever, T. M., Jenkins, D. J., Ocana, A. M., Rao, V. A., and Collier, G. R. 1988. Second-meal effect: low glycemic-index foods eaten at dinner improve subsequent breakfast glycemic response. *Am. J. Clin. Nutr.* 48:1041–1047.

Wolever, T. M. S., Bentum-Williams, A., and Jenkins, D. J. A. 1995. Physiological modulation of plasma free fatty acid comcentrations by diet: metabolic implications in nondiabetic subjects. *Diabetes Care* 18:962–970.

Wolever, T. M. S., Yang, M., Zeng, X. Y., Atkinson, F., and Brand-Miller, J. C. 2006. Food glycemic index, as given in glycemic index tables, is a significant determinant of glycemic responses elicited by composite breakfast meals. *Am. J. Clin. Nutr.* 83:1306–1312.

Wood, P. 1986. Oat β-glucan structure, location and properties. Pages 121–152 in: *Oats Chemistry and Technology.* F. H. Webster, ed. American Association of Cereal Chemists, St. Paul, MN.

Xue, Q., Newman, R. K., and Newman, C. W. 1996. Effects of heat treatment of barley starches on in vitro digestibility and glucose responses in rats. *Cereal Chem.* 73:586–592.

Yang, J.-L., Kim, Y.-H., Lee, H.-S., Lee, M.-S., and Moon, Y. K. 2003. Barley β-glucan lowers serum cholesterol based on the up-regulation of cholesterol 7α-hydroxylase activity and mRNA abundance in cholesterol-fed rats. *J. Nutr. Sci. Vitaminol.* 49:381–387.

Yokayama, W. H., and Shuo, Q. 2006. Soluble fibers prevent insulin resistance in hamsters fed high saturated fat diets. *Cereal Foods World* 51:16–18.

Yokayama, W. H., Hudson, C. A., Knuckles, B. E., Chiu, M.-C.M., Sayre, R. N., Turnlund, J. R., and Schneeman, B. O. 1997. Effect of barley β-glucan in durum wheat pasta on human glycemic response. *Cereal Chem.* 74:293–296.

Zapsalis, C., and Beck, R. A. 1985. Page 349 in: *Food Chemistry and Nutritional Biochemistry.* Wiley, New York.

Zhang, J. X., Lundin, E., Andersson, H. K., Bosaeus, I., Dahlgren, S., Hallmans, G., Stenling, R., and Åman, P. 1991. Brewer's spent grain, serum lipids, and fecal sterol excretion in human subjects with ileostomies. *J. Nutr.* 121:778–784.

9 Current Status of Global Barley Production and Utilization

BARLEY PRODUCTION

Barley is one of the seven internationally grown cereal grains, currently ranking fourth in world production behind maize, wheat, and rice and ahead of sorghum, oats, and rye (FAO 2006). Barley's rank among the major cereal grains in world production has not changed greatly in the past 15 years. Roughly, barley production in 2005 was 20% that of maize and 22% that of wheat and rice. World barley production in 2005 was approximately 138 million metric tons (MMT) produced on 56.6 million hectares (MH) (Table 9.1). Europe had the largest growing area of barley, harvesting 28.8 MH and producing 83.2 MMT in 2005, which was 60.3% of the total world barley production. North America (Canada and the United States) ranked third in area harvested (5.2 MH) and production (16.7 MMT), with 74.6% of the hectares harvested in Canada, accounting for 72.5% of the total North American production. Oceania and Africa produced similar amounts, and South America produced the least. Yield per land area harvested was greatest for North America, followed by Europe, South America, Oceania, Asia, and Africa.

Total world barley production decreased significantly over the last 25 years (Table 9.2). There was an increase in hectares harvested and tonnage produced between the early 1960s and the late 1970s, although significant decreases in production were reported over the next two decades. During the same period, barley yields per unit of area harvested increased by nearly 60% (1.46 to 2.44 metric tons per hectare), and total volume produced was increased by 38.3%. The increase in total production despite decreased harvested area may be attributed to improved genotypes and modern cultural practices, such as more effective weed control, balanced fertilization, and irrigation. It is also a possibility that those areas taken out of barley production were less productive than those that were maintained in production (D. R. Clark, personal communication).

Barley for Food and Health: Science, Technology, and Products,
By Rosemary K. Newman and C. Walter Newman
Copyright © 2008 John Wiley & Sons, Inc.

TABLE 9.1 World Barley Production in 2005: Area Harvested and Production[a]

Geographical Area	Hectares Harvested (× 1000)	Metric Tons Produced (× 1000)
Europe	28,841 (51.0)	83,202 (60.3)
Asia	12,056 (21.3)	21,302 (15.4)
North America	5,203 (9.2)	16,746 (12.1)
Canada	3,880 (74.6)[b]	12,133 (72.5)[b]
USA	1,323 (25.4)[b]	4,613 (27.5)[b]
Oceania[c]	4,789 (8.5)	10,269 (7.4)
Africa	4,897 (8.7)	4,517 (3.3)
South America	811 (1.4)	1,897 (1.4)
Total	56,597	137,933

Source: FAO (2006).
[a] Values in parentheses are percentages of the total.
[b] Percentage of North American production.
[c] Australia and New Zealand.

TABLE 9.2 Total World Barley Production in Three Periods

World	Hectares Harvested (× 1000)	Metric Tons Produced (× 1000)	Metric Tons/Hectare Harvested
1961–1965	68,071	99,716	1.465
1978–1980	84,818	167,627	1.976
1996–1998	64,300	149,800	2.317
2005	56,597	137,933	2.437

Source: FAO (1976, 1980, 1998) Year books; FAO (2006) World Crop Production.

Barley production in the United States from 2002 to 2007 is shown in Table 9.3. Total tonnage produced was highest in 2003 and 2004 and least in 2006. Yields per land area harvested in the United States were considerably higher than the average for world production and relatively consistent for the six-year period. The total land area devoted to growing barley in the United States has been reduced significantly over the past 25 to 30 years due to increased demand for other cereals, especially wheat.

It has been said with good authority that barley grows on the "frontiers of agriculture." This means that barley grows farther into the deserts, higher into the mountains, at greater latitudes, and on soils where only marginal production can be achieved from other cereal crops. Barley is our most dependable cereal crop where alkali soils, summer frost, or drought are encountered. This is not to say that barley does not respond to good fertile soil and good cultural management practices, as the best production can be achieved under ideal growing conditions with improved and locally adapted cultivars. Barley grows particularly well where the ripening season is long and cool, where the rainfall is moderate

TABLE 9.3   Barley Production in the United States, 2002–2007

Year	Hectares Harvested (× 1000)	Total Metric Tons Produced (× 1000)	Metric Tons Produced/Hectare
2002	1669	4940	2.96
2003	1913	6059	3.17
2004	1627	6091	3.74
2005	1323	4613	3.49
2006	1194	3923	3.29
2007	1420	4612	3.25

Source: National Agricultural Statistic Service (2007).

rather than excessive, and where the soil is well drained but not sandy. Barley can withstand high temperatures if the humidity is low, but it does not do so well where both heat and humidity are high. Winter production is possible at lower latitudes, and barley is more winter hardy than oats but less hardy than wheat or rye (Wiebe 1978).

The genetic versatility of barley is remarkable among grains. As the late R. F. Eslick often said, "Barley isn't barley." He referred to the wide genetic variabilities and possibilities in the barley genome. The desirable characteristics of barley for developing new products need not be limited to the existing cultivars. Breeders and geneticists have the tools to incorporate compositional properties into cultivars to meet specific needs. Indeed, the potential of transgenics holds unknown keys to future utilization of barley, as well as other grains, to feed a future hungry world.

BARLEY UTILIZATION

Historically, barley was a major dietary component of people in the areas where this grain evolved, first as a wild plant and later as a domesticated crop. Change in preference for other cereals, especially wheat, was based not on nutritional considerations but on texture, appearance, color, and perceived status of wheat-based bread products. In terms of human consumption, barley gradually became associated almost exclusively with malt and brewing. The major early use of barley for beverage production was certainly true in North America, as barley cultivation proceeded with the migration of settlers from the east coast to the midwest and on to the northwest and west coast areas. Barley production in North America during this period was concentrated near population centers, primarily to provide the raw product for malt beverage production (Weaver 1950).

Although barley is considered to be one of oldest cultivated cereal grains and was used extensively as a food in the past, it has generally been relegated to either animal feed (about 60%), malt (about 30%), and seed (about 7%), with only a small amount (about 3%) used for human food in most countries. Barley continues to be a major dietary constituent in parts of Asia and North Africa. The current use of barley for food in the United States ranks a poor third (1.5%) behind uses as animal feed and for malt/alcohol production. Barley use as food in the European Community (12 countries in 1991) was even less (0.3%) than in the United States. On the other hand, during the same time period, food was the largest use for barley in Morocco (61%), Ethiopia (79%), China (62%), and India (73%) (Kent and Evers 1994). There has been recent interest by markets in Asia for importing food barley from the United States and Canada, with special interest in waxy genotypes (USGC 2007).

Pearled barley is the most common form of barley food, recognized by most of today's modern supermarket shoppers. A multitude of barley foods, including a variety of barley soups, are possible using techniques and processing equipment traditionally used with other small-grain cereals, such as oats, rye, and wheat. Hot porridges, pilafs, pasta, snacks, and breads of many types can be made from barley. By replacing portions of white or whole-grain wheat flour with milled barley, the nutritional value and quality of many food products can be greatly enhanced. A small portion of barley malt is used in food products, principally to enhance flavor and to provide α-amylase and other enzymes to enhance baking of various breads and pastries. The major use of barley malt, however, is in the production of alcoholic beverages, especially beer. As referred to in this book, malt is essentially germinated barley. Modern malting processes are extremely complex and regulated to produce specific malt products to meet specifications of the brewing and distilling industries (Bamforth and Barclay 1993). Although alcoholic beverages are not foods in the strictest sense, their use in moderation either with or without meals adds to the total nutritional experience.

OUTLOOK FOR BARLEY FOOD

In May 2005, a workshop on "The Future of Barley" was sponsored by the American Association of Cereal Chemists International in Minneapolis, Minnesota (Anonymous 2005). Topics of the workshop covered a range of subjects, including the history of barley foods, emerging technology in barley breeding, specialized product development, health benefits, and factors affecting consumer acceptance of barley food products. Challenges were issued to barley researchers and the cereal industry to utilize the information generated on specific functional and nutritional characteristics of barley genotypes to move products to the marketplace while considering the importance of improving the acceptability of products containing barley. This workshop was viewed as a concerted effort on the part of international barley workers and the cereal industry to put barley once again at the forefront of cereals.

In May 2006, the U.S. Food and Drug Administration gave final approval of a health claim for barley, based on research demonstrating that regular consumption of barley will prevent or control cardiovascular disease by lowering blood cholesterol (FDA 2006). It is generally accepted that diets rich in whole grains provide protection against hypertension, stroke, cardiovascular disease, and type 2 diabetes. Barley and oats are both noted for their content of soluble fiber, which acts to alleviate and prevent these chronic diseases. Barley is unique among grains in that its soluble fiber extends throughout the kernel rather than being confined to the outer bran layers. Some barley varieties contain a high proportion of amylose in the starch that converts into resistant starch, a beneficial component for intestinal health.

In Chapter 7, research on barley food products is reviewed. Acceleration of these efforts in recent years indicates awareness and interest in providing consumers with barley's health benefits. Bread, a universal food, is a likely product to incorporate barley β-glucans from either the grain or in extracted form into the diet of a large population segment. To date, there has been little consumer education about barley's health benefits, and therefore little demand. Many nutritionists and food scientists that are well versed about most commonly consumed foodstuffs do not have sufficient information about barley as a food. Misconceptions and apparent transposing of nutritional information about similar grains to barley are commonly seen in nutritional recommendations. It appears that promotion of barley and new barley products by the food and cereal industries will be required to provide consumer awareness of the uniqueness, versatility, and health benefits in order to increase demand.

A compilation of consumer and producer research reports indicate that market trends in 2007 are positive for health and wellness food products (Feder 2007). Heart health is reported as a top concern for consumers, followed by general well-being, healthy fats in foods, and nutraceutical products. Weight management and obesity, closely associated with the current pandemic of diabetes and metabolic syndrome, are prime consumer concerns. Barley and whole grains in general were cited as target macro ingredients for processors focusing on heart health and other health management strategies. Dietary fiber, particularly soluble fiber, was cited as an important industry focus in the development of new food products. According to a recent report, the overall U.S. fiber market is predicted to grow to $470 million by 2011, with the soluble fiber section expected to increase by almost twice the compound annual growth rate of that of insoluble fiber (Anonymous 2007). At this time, there is active research in several parts of the world to extract, isolate, and concentrate barley β-glucans for inclusion in numerous cereal-based products. The β-glucans in barley are a proven functional food ingredient that can greatly expand the global use of whole-grain barley, processed barley, and β-glucan concentrates in numerous food products (Palmer 2006).

REFERENCES

Anonymous. 2005. The future of barley foods. *Cereal Foods World* 50:271–277.

Anonymous. 2007. Barley products to carry heart health claim. Food Navigator-USA. Published online at http://www.foodnavigator-usa.com. Accessed Oct. 20, 2007.

Bamforth, C. W., and Barclay, A. H. P. 1993. Malting technology and the uses of malt. Pages 297–354 in: *Barley: Chemistry and Technology*. A. W. MacGregor and R. S. Bhatty, eds. American Association of Cereal Chemists, St. Paul, MN.

FAO (Food and Agriculture Organization). 1976, 1980, 1998. Year books. Published online at http://www.faostat.fao.org.

FAO. 2006. World crop production. Published online at http://www.faostat.fao.org.

FDA (U.S. Food and Drug Administratin). 2006. Food labeling: health claims; soluble dietary fiber from certain foods and coronary heart disease. *Fed. Reg.* 71:29248–29250.

Feder, D. 2007. The 6 top trends in food processing. *Food Process*. Published online at http://www.foodprocessing.com/articles.

Kent, N. L., and Evers, A. D. 1994. *Kent's Technology of Cereals*, 4th ed. Elsevier Science, Oxford.

National Agricultural Statistics Service. 2007. Published online at http:/www.usda.gov/quickstats.

Palmer, S. 2006. The buzz on beta-glucans. *Food Prod. Des*. Published online at http://www.foodproductdesign.com/articles/463/463_651nutrinotes.html.

USGC (U.S. Grains Council). 2007. News release, Nov. 7. USGC, Washington, DC.

Weaver, J. C. 1950. *American Barley Production*. Burgess Publishing, Minneapolis, MN.

Wiebe, G. A. 1978. Introduction of barley into the New World. Pages 1–9, in: *Barley: Origin, Botany, Culture, Winter Hardness, Genetics, Utilization, Pests*. Agriculture Handbook 338. U.S. Department of Agriculture, Washington, DC.

10 Barley Foods: Selected Traditional Barley Recipes

INTRODUCTION

Whenever barley food is mentioned, the typical reaction is "I like barley soup." Actually, in past history as well as today, barley is appropriate for a wide variety of dishes beyond soup. Barley has the potential of being incorporated into many types of home recipes, often as a substitute for part of the original wheat, rice, or other grain. Inclusion of contemporary barley recipes is increasing in popular magazines, food columns, and various Internet food sites. The National Barley Food Council (NBFC; www.barleyfoods.org.) and various grains cookbooks, (Gelles 1989; Greene 1988; Wood 1997) are good sources of barley recipes. The recipes presented here are examples of traditional foods from regions of the world where barley is commonly used. In some cases the ingredients are very simple, reflecting sparse backgrounds and limited resources. Original preparation methods were often rudimentary, without measurements and often intended for use with primitive equipment. As much as possible, these recipes have been modified for use in modern kitchens, using estimated measurements and available ingredients and equipment. U.S. household measurements have been used, and a table of measurements equivalent to metric appears in the Appendix. These recipes are mostly from a long-standing collection by the authors, some of which were handwritten with no recorded original source. The recipes have been tested and original sources provided if available.

Barley-based traditional recipe preparation is categorized into three groups: (1) those that are prepared with raw grain, whole or ground; (2) those in which the whole or pearled grain is soaked, then drained and dried; and (3) those that are prepared with grain that has been roasted, either in the oven or in a skillet on top of the stove. The latter form is characterized by its aroma and darker color. Often, the soaked and dried or roasted barley preparations were ground

Barley for Food and Health: Science, Technology, and Products,
By Rosemary K. Newman and C. Walter Newman
Copyright © 2008 John Wiley & Sons, Inc.

and stored, and in some cases fermented, before drying and storing. This is an interesting prepreparation of convenience foods from an earlier generation.

The earliest barley recipes recorded appear to be based on ground whole meal, which was probably sifted to remove hulls. There is no mention of pearled barley until eighteenth-century reference to Scotch broth, which included whole barley, presumably in a crudely blocked form. Regions such as Tibet and Morocco utilized whole-grain hulless barley, which does not require pearling. This is still true today in those areas as well as North America and Europe. Although some cultures prefer to pearl hulless barley to obtain a small white grain to blend in with rice, the use of hulless barley is a distinct advantage for food processing.

MIDDLE EAST

Balady Bread (Pita, or Pocket Bread)

The flatbread *balady* is made throughout the eastern Mediterranean area, usually with whole wheat flour or a mixture of whole wheat and barley flours. The loaves puff up to form two layers when baking, which is convenient for eating with a filling. Indigenous Bedouin tribes living in desert areas made a similar bread without leavening, often with all barley flour, baked over an open fire on a griddle (Davidson 1999).

2 tsp dry yeast

$2\frac{1}{2}$ cups lukewarm water

$2\frac{1}{2}$ cups barley flour

3 to $3\frac{1}{2}$ cups whole wheat flour (or a blend of whole wheat and white flours)

1 Tbsp salt

1 Tbsp olive oil

In a large bowl, spread the yeast over the warm water, and stir to dissolve. Stir in about half of the flour, then stir for about 1 minute. Cover, and let the dough rest for 1 hour. Stir in the salt and olive oil, then add the remaining flour gradually until a stiff dough is reached. Turn the dough out on a floured board and knead for 8 to 10 minutes. Wash the bowl, dry, and coat with additional oil. Return the ball of dough to the bowl, cover, and allow the dough to rise until double in size. Punch the dough down and cut into 16 pieces. Roll the pieces out into circles 8 to 9 inches in diameter. Bake at 450°F for 2 to 3 minutes, or until puffed. Breads can also be baked on a hot griddle, although the expansion will be less.

Barley–Yogurt Soup

This barley–yogurt soup is made in Turkey, on the Anatolian Plateau, where barley is grown. Yogurt has been a staple in Turkey for many centuries, and barley is frequently used to make flatbreads such as yufka or pita. This recipe is from Saari and Hawtin (1977) with modifications.

1 cup pearled barley

6 cups beef or chicken broth

1 onion, chopped

$\frac{1}{4}$ cup butter

$\frac{1}{4}$ cup chopped parsley

2 Tbsp fresh or dried mint

4 cups plain yogurt

Salt and pepper to taste

Soak the barley in cold water overnight, then drain. Cook the barley in the broth until tender. Fry the onion in butter until lightly browned, then add to the barley and broth. Add the seasonings and simmer for $1\frac{1}{2}$ hours. It is desirable for the barley grains to be very soft, resulting in a thickening effect in the soup. Add the yogurt and cook over low-to-medium heat for 5 minutes, stirring constantly in one direction only. It is important that the soup not reach a boiling point once the yogurt is added, or it will curdle. Serve at once.

EASTERN EUROPE

Kasha

Kasha is a traditional dish throughout Eastern Europe, especially Russia and Poland. There are both sweet and savory versions, using barley or other available grains, such as buckwheat. A sweet barley *kasha* simply consists of barley grits cooked in water and milk and eaten with melted butter, sugar, and cinnamon. Savory *kashas* can be cooked with meat with caraway or other spices, sometimes served with grated cheese. Cast iron pots were used traditionally, but earthenware or modern baking dishes work just as well. The following recipe is of the savory type and is from Saari and Hawtin (1977). The egg coating of the barley grains provides a pleasant, chewy texture.

1 egg

1 cup pearled barley

$\frac{1}{2}$ lb fresh mushrooms or 1 oz dried mushrooms, soaked

3 to 4 Tbsp butter

2 cups water

$\frac{1}{2}$ tsp salt

2 Tbsp grated cheese

Beat the egg and stir it into the barley so that the grains are well coated, then dry in a slow oven for 30 minutes. In a large pot sauté the mushrooms in part of the butter until tender. Add the barley, water, and salt. Cover, bring to a boil, then simmer for 10 minutes. Transfer the mixture to a covered baking dish. Add sausage or other meat as desired. Bake in a moderate oven for 1 hour. Served with the remaining butter and grated cheese.

Barley–Mushroom Soup (Krupnick Polski)

This barley–mushroom soup is widely quoted as a traditional Polish dish and may be made with or without the sour cream. One version is also made with poultry giblets instead of the meat. The recipe is taken from Ochorowicz-Monatowa (1958), with modifications.

$\frac{1}{2}$ lb beef bones with some meat attached
1 cup diced vegetables, such as onion, carrots, celery, or green beans
2 dried mushrooms, soaked in warm water and sliced
3 or 4 medium potatoes, diced
$\frac{1}{2}$ cup pearled barley, cooked separately in water and drained
1 cup sour cream
Salt and pepper
1 or 2 egg yolks, beaten slightly
1 Tbsp parsley and/or dill

Place the bones, meat, mixed vegetables, and mushrooms in a pot and cover with water. Bring to a boil, then reduce the heat and simmer until the meat is tender. Add the potatoes and cook until tender. Remove the bones, trim the meat, and return the meat, together with the cooked barley, to the soup pot. Bring to a boil, then lower the heat to very low. Add the sour cream, stirring in well. Add the egg yolks gradually, stirring to prevent curdling, remove from heat after 5 minutes. Season to taste, and garnish the soup with herbs.

Goose and Barley Soup

This goose and barley soup is described as a Russian dish, although quite probably it was eaten throughout the eastern European region. The recipe is by courtesy of T. Shamliyan, University of Minnesota Medical School, Minneapolis, Minnesota.

$\frac{1}{2}$ medium goose or 1 duck
2 Tbsp cooking oil
1 onion, chopped
1 stalk celery, chopped
$\frac{3}{4}$ lb mushrooms, chopped
1 cup pearled barley
1 cup chopped tomatoes, fresh or canned

6 cups water
$\frac{1}{4}$ cup parsley
1 Tbsp dried dill
Salt and pepper

Heat the oil in a large pot. Add the goose or duck, and brown over medium heat. Cover the bird with water and simmer until tender. Remove the meat from the bird carcass and set aside. Place the onion, celery, mushrooms, and barley in a skillet with a small amount of oil, and sauté until the vegetables are soft. Return the meat to the pot with the tomatoes, sautéed vegetables, and barley. Cover and simmer 45 minutes, or until the barley is tender. Add the parsley, dill, and salt and pepper to taste. (*Author's note*: It is recommended that excess goose fat be skimmed off to meet modern tastes for low fat. Also, chicken may be substituted for goose or duck, although the flavor will not be the same.)

Meat and Barley Casserole (Miežu Putra)

This meat and barley casserole is a traditional Latvian dish, perhaps considered the national dish. It is described as a kind of vegetable/cereal porridge, usually made with barley, to which available meat may be added. (Davidson 1999). Dark rye bread is often eaten with Latvian meals. For special holiday occasions, the putra was served with a pig's head and additional sausage.

1 cup pearled barley, toasted in the oven at 325°F for 10 to 15 minutes
6 cups water
$\frac{1}{2}$ lb diced ham, bacon, or sausage
1 large onion, chopped

Toast the barley in an oven at 325°F for 10 to 15 minutes. Cook the toasted barley in the water until almost tender, then add the remaining ingredients. Continue cooking, covered, over low heat or transfer to the oven for about 1 hour.

Black Kuba (Černý Kuba)

This Czech mushroom–barley casserole is traditionally served on Christmas Eve, with cabbage or celeriac salad or pickled beets. The recipe was kindly provided by Jarka Ehrenbergerová of the Mendel Institute in Brno, Czech Republic.

1 cup dried mushrooms, soaked in water for 2 hours
1 cup pearled barley
2 Tbsp butter or cooking oil
3 cups water
1 onion, chopped

2 cloves garlic, minced

Marjoram, black pepper, cracked caraway seed, and salt (amounts to taste)

Sauté the barley in butter or cooking oil until lightly browned. Place the barley in a large pot with the water. Sauté the onion in additional oil until transparent, and add it with drained, chopped mushrooms, garlic, and seasonings to the barley pot. Continue cooking until the barley is tender. Place the mixture in a greased pan and bake at 350°F for 30 minutes.

Pork, Bean, and Barley Stew

This Yugoslavian dish is often served with pickled cucumbers or sauerkraut. Similar dishes with these ingredients are found throughout all of eastern Europe.

$\frac{1}{2}$ cup dried beans

1 lb smoked ham and/or bacon

1 cup pearled barley

1 cup potatoes, diced

1 cup parsnips, diced

$\frac{1}{2}$ cup onion, chopped

2 cloves garlic, minced

1 cup diced tomatoes

$\frac{1}{2}$ cup green pepper, diced

Soak the beans overnight or for 8 hours. Combine the soaked beans, barley, and ham or bacon. Cover with water, bring to a boil, then simmer for 1 hour. Add the remaining ingredients, and cook until all the vegetables are tender. Season to taste.

EAST AFRICA: ETHIOPIA

Injera

Injera is a fermented, spongy pancakelike bread that is eaten with meat or vegetable saucelike entrées. Fine-grained raw barley flour is used, alone or blended with other flours, such as tef. *Injera* is central to Ethiopian society. "Have you eaten *injera* today?" is a standard greeting. "He has no wat" (sauce) on his *injera* means "He's desperately poor" (Davidson 1999). This recipe is from Bekele et al. (2005), who describe several traditional Ethiopian dishes.

2 cups barley flour, or a blend of barley with wheat or tef flour

4 cups water (approximately)

2 tsp dry yeast (optional)

Blend the flour with the water to form a dough. Knead the dough until smooth, and allow it to ferment for 2 to 7 days. The fermentation time can be decreased by adding the optional yeast in the first step. After fermentation, blend the dough with enough hot water to form a smooth batter, then allow the dough to rise in a warm place. When bubbly, the batter is poured on a hot skillet or griddle to bake until brown.

SCANDINAVIA

Barley Porridge

According to Lars Munck of the Danish Royal University (Munck 1977), who provided this and the following two recipes, porridge was a major food in Old Denmark. The form of barley is often quoted as being "abraded barley grains," which probably indicates that only the hulls were removed. This would require presoaking and a long cooking period. There was probably a pot of barley cooking over a banked fire most of the time in many farmhouses. Barley porridge is described merely as abraded barley grains cooked for 2 to 3 hours in water or a mixture of water and milk. Dried fruit was sometimes added near the end of cooking.

Danish Pancakes

1 cup barley flour

2 tsp dry yeast

$\frac{3}{4}$ cup warm water

2 Tbsp honey

1 egg, beaten

1 cup plain yogurt

1 Tbsp butter, melted

Dash of cinnamon, if desired

Combine the barley flour, yeast, water, and honey. Cover and put in a warm place for $\frac{1}{2}$ hour. Add the egg, yogurt, butter, and cinnamon. Bake on a hot, greased griddle.

Fruit Soup

Fruit soups are part of the Scandinavian diet, made with available fresh or dried fruits or berries, and can be eaten at any occasion, not strictly in the form of a soup. Barley is used as a thickening agent. In the following recipe, any red juice, such as cranberry or pomegranate, may be substituted for the elderberry juice.

6 cups water

$\frac{1}{2}$ cup barley grits

$\frac{1}{2}$ cup dried pitted prunes

$\frac{1}{2}$ cup raisins

$\frac{1}{2}$ to 1 cup elderberry juice

$\frac{1}{2}$ cup sugar

The barley grits are soaked in water overnight, then cooked over low heat until tender, which may take 2 hours or more. The fruit is added in the last hour of cooking, and sugar and fruit juice are added just before serving. The soup is eaten hot or cold, and if available, cream is sometimes poured on top.

Barley Sausage (Pölsa)

This barley sausage dish was made in rural Sweden after the annual slaughtering day and was traditionally eaten with pickled beets. The original recipe may be modernized by using about 1 pound of ground lean pork in place of the heart, cooking it in $1\frac{1}{2}$ quarts of water with the barley, then blending the mixture in a food processor until fairly smooth but not puréed. This sausage has a form not unlike that of scrapple, a traditional Pennsylvania Dutch dish.

1 pork or beef heart

1 Tbsp salt

4 cups pearled barley

2 large onions, chopped

2 Tbsp lard or cooking oil

2 Tbsp molasses

White pepper

Dried or fresh marjoram

Clean the heart of tubes and fibers, score with a sharp fork, and cover with cold water to soak for 2 hours. Drain, cover with salted boiling water in a large pot and cook slowly until tender. Meanwhile, soak the barley in cold water. Remove the heart from the pot, and cook the drained barley in the broth. When the barley is tender, grind the meat, add the barley, mix, and grind again. Sauté the onions in the lard (or cooking oil), and add to the meat mixture. Add the molasses, pepper, and marjoram to flavor as desired. Press the sausage into a flat pan and chill. Slice the sausage and sauté before serving.

Swedish Flatbread (Kornmjölsbröd)

This swedish flatbread recipe was developed by the authors after spending a sabbatical in Uppsala, Sweden in 1982. At that time barley flour (*kornmjöl*) and barley bread were available in grocery stores. This was the beginning of an enduring fascination with barley recipes. Barley flatbread has a long history in Scandinavia, particularly in rural areas.

5 tsp dry yeast

3 Tbsp warm water

4 tsp butter

2 cups whole milk

2 tsp salt

3 Tbsp molasses or corn syrup

$\frac{1}{2}$ tsp caraway seeds, crushed (optional)

3 cups barley flour

2 cups white wheat flour

Dissolve the yeast in the warm water. Heat the milk and butter to 110°F, then place in a large bowl. Add the yeast mixture, salt, molasses, caraway, and barley flour. Mix well. Slowly fold in the white wheat flour to form a soft dough. Cover the dough and let rise for 30 minutes. Turn the dough out on a board and knead for 5 minutes. Cut into 16 pieces and roll out into circles about 5 inches in diameter. Pierce the tops of the loaves with a fork. Bake at 500°F on a greased sheet for about 7 minutes. The loaves should be puffy and slightly browned. Cool under a towel.

NORTHEASTERN EUROPE: FINLAND

Talkkuna

Talkkuna is a typical food throughout northeastern Europe, with variations in different regions. (This and the following Finnish recipe are courtesy of Hannu Ahokas, Jokioinen, Finland, who transcribed the recipes given to him by his mother, who was born in 1922.)

Talkkuna is a type of precooked, dried grain used as a convenience food for breakfast or as a starchy food with soup or meat. Barley grains, often mixed with oats and/or peas, are soaked for several hours in hot water. When the grains are swollen, they are drained and spread to dry in an oven with very low heat, so as to prevent burning. The grains are then ground to a very fine meal and stored in a dry place. *Talkkuna* is generally eaten with sour milk or yogurt. Sugar and fruit are sometimes added. *Talkkuna* can also be made into a porridge and eaten with butter on top.

Barley Baked on Leaves (Ohralehikäiset)

Half a head of fresh cabbage

1 cup pearled barley

2 to 3 cups milk or a mixture of milk and cream

1 Tbsp sugar

1 tsp salt

Spread layers of fresh cabbage leaves in a shallow pan. Combine the pearled barley, salt, sugar, and milk and spread in a $\frac{1}{2}$-inch layer on the cabbage leaves. Bake at 350°F until the barley is soft. (*Note by Hannu Ahokas*: In the ancient Finnish tradition, wild leaves of the white water lily were used to bake *lehikäiset*.)

Finnish Barley Pudding

This barley pudding dish can be eaten as a starchy alternative to potatoes, or can be sweetened and eaten as a dessert with cream or milk. The recipe is common in collections of international foods.

$1\frac{1}{2}$ cups pearled barley

6 cups hot milk

1 tsp salt

$\frac{1}{3}$ cup butter

Optional for hot side dish: pepper to taste, or for a dessert: $\frac{1}{2}$ to $\frac{3}{4}$ cup sugar and 1 tsp cinnamon if made as a dessert.

Soak the barley overnight, then bring to a boil in the same water. Add salt and cook at a simmer until the barley begins to swell up. Gradually add the hot milk and pepper, if used. Continue to simmer for 30 minutes, stirring frequently. Transfer to a buttered baking dish, dot with butter and bake in a low oven (250°F), about 2 hours, or until golden brown. If sweetened, add $\frac{1}{2}$ to $\frac{3}{4}$ cup sugar and cinnamon before baking.

GREAT BRITAIN

Orkney Bere Bannocks

Bannocks are a type of flat bread from Scotland, particularly the Orkney Islands, where a type of barley known as Bere has been used as food for centuries. Restaurants in Orkney regularly serve Bere bannocks, and the Barony Mill, powered by water, still grinds Bere into whole-grain meal. This recipe is from the Barony Mill Web site: www.birsay.org.uk/baronymill.htm. "The bannocks o' barley, bannocks o' bere meal, bannocks o' barley, Here's to the Hielandman's bannocks o'barley"—Old Scottish song

1 cup all-purpose flour

1 cup Bere meal (or any whole-grain barley flour)

1 tsp baking soda

$\frac{1}{4}$ tsp salt

$\frac{1}{4}$ tsp cream of tartar

Buttermilk (about $\frac{3}{4}$ cup)

Place the dry ingredients in a bowl. Stir in the buttermilk, sufficient to make a soft dough. Roll out lightly on a floured board to $\frac{1}{2}$-inch thickness. Cut into rounds and bake on a hot griddle until lightly brown on both sides. These are best eaten warm, with butter and jam or marmalade.

Scotch Broth

Scotch broth really means *barley broth*, and has a long history in Great Britain. Lamb or beef was used originally, together with available garden vegetables. Samuel Johnson, the eighteenth-century lexicographer and poet, was reported to have eaten Scotch broth when he visited James Boswell in Aberdeen in 1786. "Dr. Johnson ate several plates of broth with barley and peas, and seemed very fond of the dish. I said 'You never ate it before?' Dr. Johnson—'No, sir, but I don't care how soon I eat it again'" (Boswell 1786).

$\frac{1}{2}$ lb boneless lamb meat, diced

1 tsp salt

6 cups water

1 leek, chopped

1 carrot, diced

1 turnip, diced

2 cups cabbage, sliced

$\frac{1}{3}$ cup green split peas

$\frac{1}{2}$ cup pearled barley

Place the meat in a pot with the water and salt. Bring to a boil and simmer 1 hour. Add the vegetables, peas, and barley. Simmer for an additional hour. Season to taste with pepper, additional salt, and other herbs, as desired.

Barley Water

Barley water has been known in history as a therapeutic aid for various ailments as well as for prevention of heat exhaustion and dehydration. Caution should be used in infant feeding to ensure that milk or formula is not limited, depriving infants of needed calories, protein, and other nutrients.

$\frac{1}{2}$ cup barley (hulless or pearled)

2 to 3 qt water

Combine the barley and water, bring to a boil, cover, lower the heat, and simmer overnight or 8 hours. (A crockpot may also be used.) Strain the barley grains from the water. Sugar, another sweetener, or lemon juice may be added to the barley water. Depending on the variety of barley used, the barley water may have a slight color, sometimes pink, due to natural pigments in the grain. If hulless barley is used, the cooked grain may be used in various foods. For therapeutic purposes, such as intestinal disorders or mouth sores during chemotherapy, the barley water may be sipped continually throughout the day.

ASIA

Barley Mixed-Bob

Bob in Korean generally means cooked rice. Accordingly, *barley bob* or *mixed-bob* refers to a cooked mixture of rice and precooked barley. The barley was precooked because it took longer to cook to tenderness. The word *bori* means barley, and 100% cooked barley is called *kkong bori bob* (B.-K. Baik, personal communication). Historically, barley was eaten in Korea in times of small crop yields, and by people of lower economic means. Early in the twentieth century, the Korean government required that barley be mixed with rice, however, as rice production increased, the regular consumption of barley was abandoned (Ju 1979). Recently, there has been a renewal of interest in the barley–rice combination, for health reasons.

Equal parts of pre-partially cooked pearled barley and white rice are cooked in water until tender. Salt is used in the cooking water as desired. [*Authors' note*: Quick barley (Quaker Company) and white rice require about the same time for cooking, so may be combined for cooking. Alternatively, raw pearled barley and brown rice may be cooked together.]

In many Asian cultures, white polished rice is preferred, and the current practice is to pearl barley to a degree where the grain appears almost white, then to split the grains. The resulting half-grains are very similar to the white rice, and are more acceptable to consumers, whether eaten as is or mixed with white rice. In addition, the cooking times are similar, eliminating the need for precooking the barley.

Tsampa

Tsampa is a traditional Tibetan dish, which is still an important part of the modern diet, associated with celebrations and rituals. The mystique of Tibet for travelers includes the proper method of eating tsampa, and has been described and pictured on various Internet Web sites, such as http://www.tsampa.org/Tibetan/theory_ and_practice. Chapter 1 provides more information on *tsampa* preparation.

Hulless barley grains are oven-roasted at about 325°F until slightly brown. After cooling, the grain is ground in a grain mill into flour, which can be stored for future use. The *tsampa* flour is traditionally combined with hot, black butter tea called *po cha* in a small bowl. The tea is made by combining hot black tea with salt, yak butter and milk, then churning, shaking, or blending this mixture for 2 or 3 minutes. If yak butter is not available, regular butter may be used, but purists claim that the dish is not the same. There are various methods of consuming the mixture. When the tea is added on top of the *tsampa*, it may be drunk partially, leaving a doughy mixture in the bottom. This ball of dough is then manipulated with the fingers until it forms a ball, which is then eaten.

Tsampa is eaten in many different ways, in addition to mixing it with tea. It is sometimes made into cakes, mixed with butter or grated cheese, perhaps with sugar. It is also made into various porridges or soup, sometimes with meat (Tashi 2005).

SUMMARY

As reviewed in Chapter 1, barley has a long and varied history in human civilization. In many cases, barley was the only or the most abundant cereal grain available. The recipes selected for inclusion here represent a variety of cultures and food styles. The diversity of preparation of the raw grains gives a suggestion of the ingenuity of early people in preparing palatable, filling meals with limited resources. Theirs was not a question of nutritional value or health benefits, but of satisfying appetites with available grain supplies. This small collection will hopefully encourage the reader to use creativity in using barley in the modern food world.

REFERENCES

Bekele, B., Alemayehu, F., and Lakew, B. 2005. Food barley in Ethiopia. Pages 53–82 in: *Food Barley—Importance, Uses and Local Knowledge: Proc. International Workshop on Food Barley Improvement*. S. Grando and H. G. Macpherson, eds. ICARDA, Aleppo, Syria.

Boswell, J. 1786. *Journal of a Tour to the Hebrides with Samuel Johnson.*

Davidson, A. 1999. *The Oxford Companion to Food*. Oxford University Press, New York.

Gelles, C. 1989. *The Complete Whole Grain Cookbook*. Donald I. Fine, New York.

Greene, B. 1988. *The Grains Cookbook*. Workman Publishing, New York.

Ju, J. S. 1979. A study on some nutritional effects for feeding barley and rice mixed diets on rats. Pages 65–71 in: *Proc. Joint Barley Utilization Seminar*. Korea Science and Engineering Foundation, Suweon, Korea.

Munck, L. 1977. Barley as food in old Scandinavia, especially Denmark. Pages 386–393 in: *Proc. 4th Regional Winter Cereal Workshop: Barley*, vol. II. ICARDA, Aleppo, Syria.

NBFC. National Barley Food Council. Spokane, Washington.

Ochorowicz-Monatowa, M. 1958. *Polish Cookery*. Crown Publishers, New York.

Saari, K., and Hawtin, L. 1977. Back to Barley: recipes for the world's oldest food crop. Presented at the 4th Regional Winter Cereal Workshop: Barley, Amman, Jordan.

Tashi, N. 2005. Food preparation from hull-less barley in Tibet. Pages 115–120 in: *Food Barley—Importance, Uses, and Local Knowledge: Proc. International Workshop on Food Barley Improvement*. S. Grando, and H. G. Macpherson, eds. ICARDA, Aleppo, Syria.

Wood, R. 1997. *The Splendid Grain*. William Morrow, New York.

APPENDIX 1
Glossary: Botany and Plants

Allele: One of a number of alternative forms of a gene.

Anther: The part of the flower that produces the pollen.

Anthesis: Stage or period at which a flower bud opens.

Awn: Stiff, bristlelike projection from the tip or back of the lemma or glumes in grasses.

Caryopsis: Small, dry, one-sided fruit in which the ovary wall remains joined with the seed in a single grain.

Coleoptile: Protective sheath surrounding the plumule (germinating shoot) of some monocotyledonous plants, such as grasses.

Coleorhiza: Protective sheath surrounding the radicle (germinating root) of some monocotyledons, such as grasses.

Cotyledon: The first leaf or leaves of a seed plant, found in the embryo, which may form the first photosynthetic leaves or may remain belowground.

Diploid: Organism whose cells have two sets of chromosomes of the basic genetic complement.

Epicotyl: The stemlike axis of the young plant embryo above the cotyledons, terminating in an apical meristem and sometimes bearing one or more young leaves.

Floret: One of the small individual flowers of a crowded infloresance; individual flower of grasses.

Glume: Dry chaffy bract, present in a pair at the base of spikelet in grasses.

Haploid: Cells having one set of chromosomes representing the basic genetic complement of the species.

Homozygous: Diploid organism that has inherited the same allele of any particular gene from both parents.

Hordeum vulgare **L.:** Domesticated (cultivated) barley.

Barley for Food and Health: Science, Technology, and Products,
By Rosemary K. Newman and C. Walter Newman
Copyright © 2008 John Wiley & Sons, Inc.

***Hordeum vulgare spontaneum* C. Koch.:** Wild barley, two-rowed, considered to be the most recent and immediate ancestor of cultivate barley, *Hordenum vulgare* L.

Hull: The husk, or outer covering of a kernel.

Hulled barley: Barley having the hull securely attached to the pericarp.

Hulless barley: Barley not having the hull attached to the pericarp, which is mostly removed in the harvesting process.

Hypocotyl: That portion of stem below cotyledons in plant embryo which eventually bears roots.

Inflorescence: Flower head in flowering plants.

Internode: The part of a stem between two nodes.

Leaf sheath: The lower part of a leaf enclosing the stem.

Lemma: The lower of the two bracts enclosing a floret (individual flower) in grasses.

Main shoot: The primary shoot which emerges first from the soil, where tillers originate.

Meiosis: Nuclear division that results in daughter nuclei, each containing half the number of chromosomes of the parent.

Meristem: Plant tissue capable of undergoing mitosis, giving rise to new cells and tissues.

Mitosis: Nuclear division in eukaryotic cells.

Node (joint): Region on the stem where leaves are attached.

Palea: The upper of two bracts enclosing a floret (individual flower) in grasses.

Peduncle: Stalk of a flower head.

Pericarp: The layers of a fruit or seed that develop from the ovary wall, comprising an outer skin.

Rachilla: Axis bearing the florets in a grass spikelet; small or secondary rachis.

Rachis: The shaft; a stalk or axis.

Radical: Arising from a root close to the ground, as basal leaves and flower stems.

Radicle: Embryonic plant root.

Scutellum: Development of part of the cotyledon that separates the embryo from the endosperm in the seed.

Seminal root: The first root formed, developed from the radicle of the seed.

Sessile: Sitting directly on a base, without a stalk or pedicel.

Spike: The barley head, ear.

Spikelet: The flower of a grass, consisting of a pair of glumes and one or more enclosed florets.

Tiller: Shoot originating from the base of the plant, flower stem of a grass.

APPENDIX 2
Glossary: Food and Nutrition

Atherosclerosis: Thickening and loss of elasticity of the arterial walls due to accumulation of plaque.

Cholecystokinin: Hormone secreted by the intestine which stimulates secretion of bile and pancreatic enzymes.

Cytokine: Nonantibody protein which acts as an intercellular mediator, such as causing an immune response.

Dextrose equivalent: The extent of hydrolysis of starch into glucose (dextrose).

Fermentation: Breakdown of indigestible carbohydrates in the large intestine by bacteria; also enzymatic conversion of foods and yeasts.

Free fatty acids: Plasma fatty acids that are not in the form of glycerol esters.

Gelatinization: Disruption of starch molecules causing irreversible changes in properties, such as granule swelling.

Glucagon: A hormone secreted by beta cells of the pancreas in response to elevation of blood glucose.

Glycemic index: System of classifying foods according to the degree of blood glucose increment following consumption, compared to either white bread or glucose.

Glycemic response: Increase in blood glucose following consumption of a food or beverage.

HDL cholesterol: High density lipoproteins that promotes transport of cholesterol to the liver for excretion.

Hedonic: "Pertaining to pleasure." Taste testing scale for the degree of liking a product.

Hydrocolloids: Polysaccharides that form viscous solutions or dispersions in water.

Hypercholesterolemia: Level of total or LDL cholesterol in the blood higher than normal range.

Ileostomy: Surgical creation of an opening into the ileum, usually be establishing a stoma (opening) in the abdominal wall.

Barley for Food and Health: Science, Technology, and Products,
By Rosemary K. Newman and C. Walter Newman
Copyright © 2008 John Wiley & Sons, Inc.

Incretin: Gastrointestinal hormone that increases insulin release from the pancreas, and may also slow gastric emptying.

Insulin: A hormone secreted by alpha cells of the pancreas in response to low blood sugar.

LDL cholesterol: Low-density lipoprotein carrying cholesterol in the blood (often called "bad cholesterol").

Nonstarch polysaccharides: Carbohydrates that are not digested by human enzymes; generally considered dietary fiber components.

Organoleptic: Responding to sensory stimuli; sensory testing of food products.

Pasting: Phenomenon following gelatinization in the dissolution of starch, ending in the disruption of granules.

Phenols, polyphenols: Flavonoid compounds with phenyl ring structures; plant pigments found in the seed.

Postprandial: After a meal, absorptive state.

Reducing sugars: Sugars containing a hemiacetyl group that brings about reduction and is oxidized in the process.

Rheology: The change in form and flow of matter, involving elasticity, viscosity, and plasticity.

Retrogradation: After heating, followed by rapid cooling or freezing of starch, especially amylose, causing aggregation of molecules.

Satiety: The feeling of having had enough to eat, being satisfied.

APPENDIX 3
Glossary: Barley Terms

The terms in this appendix are adapted from AACC Approved Method 55–99 (2000), with permission from the American Association of Cereal Chemists, St Paul, Minnesota.

Barley: The fourth major world cereal, belonging to the family Poacea, the tribe Triticeae, and the genus *Hordeum*. All cultivated barley belongs to *Hordeum vulgare*. Cultivated barley can have either the two- or six-rowed head type upon which either hulled or hulless seed develop.

Barley bran: Product of roller or separation milling of dehulled or hulless barley, which includes the outer covering such as pericarp (nucellar epidermis), testa (seed coat), and aleurone and subaleurone layers. Should be free of hulls. May also be a combination of milled bran and shorts fractions.

Barley brewers' grain: By-product from the malting and brewing processes, consisting principally of the hull, germ, oil, protein, and unfermentable fiber. Also known as brewers' spent grain.

Barley flakes: Dehulled, pearl, or hulless barley that may be enzyme deactivated and/or tempered followed by the process of being rolled, dried, and cooled. Flake thickness can be varied depending on the original grain form and degree of rolling pressure.

Barley flour: Produced by dry milling of barley, consisting principally of endosperm tissue. Extraction rate may vary in flour produced from roller-separation milling.

Barley germ: The germ consists of the embryonic axis and the scutellum, usually obtained by separation from barley bran or barley pearlings.

Barley grits: Dehulled, pearl, or hulless barley that has been cut into small pieces. Also known as barley bits.

Barley hulls: Outer coverings of the barley kernel, consisting of the two flowering glumes, lemma and palea, which may or may not adhere to the pericarp. Also known as husks.

Barley for Food and Health: Science, Technology, and Products,
By Rosemary K. Newman and C. Walter Newman
Copyright © 2008 John Wiley & Sons, Inc.

Barley malt: Product from soaking or steeping the whole barley kernel followed by germination and drying (kilning) in a controlled environment.

Barley pearlings: The residue from the pearling process consisting of hulls, pericarp, testa, aleurone, subaleurone, germ, and various percentages of the endosperm, depending on the degree of pearling. The hulls are not part of the pearlings when dehulled or hulless barley is pearled.

Barley shorts: A fraction of the barley milling process, separate from bran and flour.

Dehulled barley: Hulled barley from which the hulls have been removed by a physical process.

High-amylose barley: Barley having the homozygous recessive gene *amol*, which increases the percentage of amylose in the starch up to 40 to 70% compared to the normal 25 to 30% amylose.

Hulled barley: Barley in which the two flowering glumes, lemma and palea, adhere to the seed. Also known as covered barley.

Hulless barley: Barley having the homozygous recessive gene *nud*, which prevents the hulls (flowering glumes) from adhering to the seed. Also known as naked barley.

Pearl barley: A barley product in which the hulls, pericarp, testa, germ, and part of the outer endosperm are removed by an abrasive scouring process. Coarse, medium, or fine pearl can be produced by increasing the amount of abrasion. Other terminology of pearl barley may include blocked, pot, or scotch barley.

Waxy barley: Barley having the homozygous recessive gene *wax*, which produces starch that is principally or completely amylopectin.

Whole-grain barley products: Flour, grits, flakes, or other products made from barley that includes the bran, germ, and endosperm. Whole-grain barley flour may also be known as whole barley meal.

APPENDIX 4
Equivalent Weights and Measures

U.S. Measure	Metric
Volume	
1 tsp (t)	5 mL
1 Tbsp (T)	15 mL
$\frac{1}{4}$ cup (4 T)	60 mL
$\frac{1}{2}$ cup (8 T)	120 mL
1 cup	240 mL
4 cups (1 qt)	950 mL
1 qt	0.95 L
4 qt (1 gal)	3.79 L
Weight	
1 oz	28 g[a]
4 oz ($\frac{1}{4}$ lb)	114 g
8 oz ($\frac{1}{2}$ lb)	227 g
1 lb (16 oz)	454 g
Conversion for Oven Temperatures	
150°F	66°C
200°F	93°C
300°F	149°C
325°F	163°C
350°F	177°C
400°F	205°C
450°F	232°C
500°F	260°C

[a] Base value = 28.35 g.

Barley for Food and Health: Science, Technology, and Products,
By Rosemary K. Newman and C. Walter Newman
Copyright © 2008 John Wiley & Sons, Inc.

APPENDIX 5
Sources of Barley and Barley Products

Alexander Company
P.O. Box 235
Bancroft, ID 83217
Ph: 206-648-7770
Contact: Wade Clark
E-mail: alexco@dcdi.net
(Wholesale barley flour and pearled barley)

Arrowhead Mills/Hain-Celestial
110 South Lanton
Hereford, TX 79045
Ph: 806-364-0730
Web: http://www.arrowheadmills.com
(Sell barley flour and pearl barley to distributors and online)

BGLife™ Barley
8 West Park St., Ste 210
Butte, MT 59701
Ph: 888-238-2458 or 701-219-3275
E-mail: customerservice@bglifebarley.com
(Retail online or wholesale barley flakes)

Barry Farm Foods
20086 Mudsock Rd.
Wapakoneta, OH 45895
Ph: 419-228-4640
E-mail: order@barryfarm.com

Barley for Food and Health: Science, Technology, and Products,
By Rosemary K. Newman and C. Walter Newman
Copyright © 2008 John Wiley & Sons, Inc.

Web: http://www.barryfarm.com
(Wholesale/retail organic barley flour, malt, malt syrup, barley flakes, and pearled)

Bob's Red Mill Natural Foods, Inc.
5209 SE International Way
Milwaukie, OR 97222
Ph. 100-349-2173
Web: http://www.bobsredmill.com
(Wholesale/retail whole hulless, pearled, and barley flour)

Briess Malt & Ingedients Co.
625 South Irish Road, P.O. Box 229
Chilton, WI 53014
Ph: 920-849-7711
E-mail: info@briess.com
(Wholesale malted barley flours, flakes, grits, and roasted barley; 50 lb minimum)

Can-Oat Milling, Inc.
1 Can-Oat Drive
Portage La Prairie
MB R1 N 3 W1 Canada
(Wholesale barley flour, flakes, pot, and pearled barley)

Cargill, Inc.
15407 Mcginty Rd. W.
Wayzata, MN 55391
Ph. 952-742-7575
Web: http://www.cargill.com
(Wholesale Barliv, β-glucan concentrate, and barley malt)

Circle S Seeds of Montana, Inc.
P.O. Box 130
Three Forks, MT 59752
Ph: 406-285-3269
Contact: Steve McDonnell
E-mail: circles@theglobal.net

ConAgra Mills
11 ConAgra Drive
Omaha, NE 68102

Ph: 402-595-4000
Web: http://www.conagramills.com
(Wholesale Sustagrain high-β-glucan barley in whole, steel-cut, flaked, and flour forms)

GraceLinc Ltd.
Christ Church, New Zealand
+64-3-325-9683
E-mail: john.morgan@glucagel.com
Web: http://www.glucagel.com
(Market Glucagel and β-glucan concentrate)

Grain Millers, Inc.
9531 W. 78th St. No. 400
Eden Prairie, MN 55344
Ph: 800-232-6287
Or: 315 Madison St.
Eugene, OR 97402
Ph: 800-443-8972
Web: http://www.grainmillers.com
(Wholesale organic and conventional barley flakes and steamed barley flour)

Hamilton's Barley
R.R. 2, Olds, Alberta
T4H 1P3 Canada
Ph: 403-556-8493
Web: http://www.hamiltonsbarley.com
(Wholesale barley flour)

Hesco, Inc.
500 19th St. SW, Box 815
Watertown, SD 57201-0815
Contact: Jeff McGinley
Ph: 605-884-1100
E-mail: hescoinc@hesco-inc.com
Web: http://www.hesco-inc.com
(Wholesale pearl barley, barley grits, and barley flour)

Honeyville Food Products
11600 Dayton Drive

Rancho Cucamonga, CA 91730
Ph: 888-810-3212
Web: http://www.honeyvillegrain.com
(Wholesale/retail pearled barley, flour, and flakes)

Indian Harvest Special Foods, Inc.
P.O. Box 428
Bemidji, MN 56619
Ph. 800-346-7032
(Wholesale black barley only)

King Arthur Flour
The Baker's Store
135 Rt. 5 South
Norwich, VT 05055
Ph: 800-827-6836
Web: http://www.kingarthurflour.com
(Wholesale/retail barley flour, flakes, and steel-cut barley)

Minnesota Grain, Inc.
1380 Corporate Ctr. Curve, Ste. 105
Eagan, MN 55121
Ph: 651-681-1460
E-mail: info@mngrain.com
Web: http://www.mngrain.com
(Wholesale barley flour, pearl barley, and hulless barley flakes; 50 lb minimum)

Montana Milling
2123 Vaughn Rd.
Great Falls, MT 59404
Ph: 406-771-9229
Web: http://www.montanamilling.com
(Wholesale/retail organic and commercial whole barley, barley flour, barley flakes, and grits)

Natural Way Mills, Inc.
24509 390th St., NE
Middle River, MN 56737
Contact: Ray Juhl
Ph: 218-222-3677

Web: http://www.naturalwaymills.com
(Wholesale/retail organic hulled barley, barley flour, and pearled barley)

NuWorld Nutrition
816 6th Avenue NE
Perham, MN 56573
Contact: Wally Corum
Ph: 800-950-3186
E-mail: nuworld@eot.com
Web: http://www.eot.com/~nuworld/index.html
(Retail mail-order business selling waxy hulless barley flakes)

Palouse Grain Growers, Inc.
110 West Main St., Box 118
Palouse, WA 99161
Contact: Bruce Baldwin
Ph: 800-322-1621
E-mail: grain@palouse.com
Web: http://www.users.palouse.com/grain
(Wholesale, bulk grain coop; processes pearl barley; 25 lb minimum)

Roman Meal Milling Company
4014 15th Avenue NW
Fargo, ND 58102
Contact: Bill Fletcher
Ph: 701-282-9656
E-ml: bfletcher@romanmealmilling.com
Web: http://www.romanmealmilling,com
(Wholesale specialty flour, blends, and flakes)

Roxdale Foods Ltd.
Private Box 302-860
North Harbour
Auckland, New Zealand
Ph: 64 9 415 1135
Email: hamishd@roxdale.co.nz
Web: http://www.roxdale.co.nz
(Market Cerogen and β-glucan concentrate)

Quaker Oats Company
321 N. Clark St.
Chicago, IL 60610
Ph: 312-821-1000
Web: http://www.quakeroats.com
(Market pearl barley and quick barley in stores under Scotch and Mother's Cereal brand names; sells online also)

Western Trails Inc/Cowboy Foods
313 W. Valentine St.
Glendive, MT 59230
Contact: Rachel Williams
Ph: 406-377-4284
E-mail: info@westerntrailsfood.com
Web: http://www.westerntrailsfood.com
(Wholesale/retail barley flours, hulless barley, soup mixes, and bread mixes)

APPENDIX 6
Barley Resource Organizations

Alberta Barley Commission
3601 A 21st St. NE
Calgary, Alberta
T2E 6 T5 Canada
Ph: 403-291-9111
E-mail: barleyinfo@albertabarley.com
Web: http://www.albertabarley.com

American Malting Barley Association
740 N. Plankinton Ave., Ste. 830
Milwaukee, WI 53203
Ph: 414-272-4640
E-mail: info@AMBAinc.org
Web: http://www.AMBAinc.org

Barley CAP (Coordinated Agricultural Project)
University of Minnesota
Department of Agronomy
Dr. Gary Muehlbauer, Director
1991 Buford Circle, 411 Borlaug Hall
St. Paul, MN 55103-6026
Ph: 612– 6256228
Web: http://www.barleycap.dfans.umn.edu

CSIRO Australia
CSIRO Enquiries
Bag 10, Clayton South
VIC 3169 Australia

Barley for Food and Health: Science, Technology, and Products,
By Rosemary K. Newman and C. Walter Newman
Copyright © 2008 John Wiley & Sons, Inc.

E-mail:Tony.Steeper@csiro.au
Web: http://www.csiro.au/science/pshh.html

ICARDA
International Center for Agricultural Research in Dry Areas–Genebank
ICARDA, P.O. Box 5466
Aleppo, Syria
E-mail: J. Valkoun@CGIAR.ORG

Idaho Barley Commission
821 W. State St.
Boise, ID 83702
Ph: 208-334-2090
E-mail: kolson@idahobarley.org
Web: http://www.idahobarley.org

Minnesota Barley Research and Promotion Council
2601 Wheat Drive
Red Lake Falls, MN 56750
Ph: 218-253-4311
Web: http://www.mda.state.mn.us/food/business/promotioncouncils.htm

Montana Wheat and Barley Committee
P.O. Box 3024
Great Falls, MT 59403-3024
Ph: 406-761-7732
E-mail: wbc@mt.gov
Web: http://www.wbc.agr.mt.gov

National Barley Food Council
907 W. Riverside Ave.
Spokane, WA 99201
Ph: 509-456-4400
E-mail: mary@wagrains.com
Web: http://www.barleyfoods.org

National Barley Growers Association
505 40th St. SW, Ste. E
Fargo, ND 58103
Ph: 701-239-7200
E-mail: Steven.edwardson@ndbarley.net

National Barley Improvement Committee
740 N. Plankton Ave., Ste. 830
Milwaukee, WI
Ph: 414-272-4640
E-mail: info@AMBAinc.org
Web: http://www.AMBAinc.org

Nordic Barley Gene Bank
P.O. Box 41
SE-23053 Alnarp, Sweden
Web: http://www.nordgen.org

North Dakota Barley Council
505 40th St. SW
Fargo, ND 58103
Ph: 701-239-7200
E-mail: ndbarley@ndbarley.net

Northern Crops Institute
Bolley Drive, NDSU
Fargo, ND 58105
Ph: 701-231-7736
E-mail: nci@ndsu.edu
Web: http://www.northern-crops.com

Oregon Grains Commission
115 SE 8th, P.O. Box 1086
Pendleton, OR 97801
Ph: 541-276-4609
Web: http://www.owgl.org

Oregon State Univ. Barley Project
Crop and Science Department
OSU, Corvallis, OR 97331
Web: http://www.barleyworld.org

Phoenix AGRI Research/Westbred LLC
717 14th St. S.
Fargo, ND 58103
Drs. Chris Fastnaught and Greg Fox, Consultants

Ph: 701-293-5146
E-mail: drgfox@hotmail.com

Plant Gene Resources of Canada
Agriculture and Agri-food Canada
Saskatoon Research Center
107 Science Place
Saskatoon, Saskatchewan
S7 N 0X2 Canada
Web: http://www.pgrc3.agr.ca

Svalöf-Weibull AB
SE-268.81 Svalöf, Sweden
E-mail: info@swseed.com
Web: http://www.swseed.com

University of Saskatchewan Crop Development Centre
51 Campus Drive
Saskatoon, Saskatchewan
S7 N 3 R2 Canada
Ph: 306-966-4343

USDA National Small Grains Collection
USDA, ARS, Pacific West Area
1691 S. 2700 W.
Aberdeen, ID 83210
E-mail: Mike.Bonman@ars.usda.gov
Web: http://www.ars.usda.gov

U.S. Grains Council
1400 K St. NW, Ste. 1200
Washington, DC 20005
Ph: 202-789-0789
E-mail: grains@grains.org
Web: http://www.grains.org

Washington Grain Alliance
907 W. Riverside Ave.
Spokane, WA 99201-1006
Ph: 509-456-4400

E-mail: wga@wagrains.com
Web: http://www.washingtongrainalliance.com

WestBred LLC
81 Timberline Drive
Bozeman, MT 59718
E-mail: info@westbred.com
Web: http://www.westbred.com

Western Grain Research Foundation
214-111 Research Drive
Saskatoon, Saskatchewan
S7 N 3 R2 Canada
Ph: 306-975-0060
E-mail: info@westerngrains.com

INDEX

Barley for Food and Health: Science, Technology, and Products,
By Rosemary K. Newman and C. Walter Newman
Copyright © 2008 John Wiley & Sons, Inc.

Printed in the United States
By Bookmasters